C. LAIRD

CHROMOSOMES TODAY

VOLUME 9

Proceedings of the Ninth International Chromosome Conference held in Marseille, France, 18–21 June 1986

Editors
A. STAHL
J.M. LUCIANI
A. M. VAGNER-CAPODANO
Faculté de Médecine, Marseille

Series Editor
M. D. BENNETT *Cambridge*

London
ALLEN & UNWIN
Boston Sydney Wellington

This book was typeset, proofed and passed for press by the authors and editors.

Allen & Unwin, the academic imprint of

Unwin Hyman Ltd

PO Box 18, Park Lane, Hemel Hempstead, Herts HP2 4TE, UK
40 Museum Street, London WC1A 1LU, UK
37/39 Queen Elizabeth Street, London SE1 2QB

Allen & Unwin Inc.,
8 Winchester Place, Winchester, Mass. 01890, USA

Allen & Unwin (Australia) Ltd,
8 Napier Street, North Sydney, NSW 2060, Australia

Allen & Unwin (New Zealand) Ltd in association with the Port Nicholson Press Ltd,
60 Cambridge Terrace, Wellington, New Zealand

First published in 1987

ISSN 0069–3944

Printed in Great Britain by St. Edmondsbury Press, Suffolk

Contents

NUCLEOLUS ORGANISER

MARSEILLE CHROMOSOME CONFERENCE

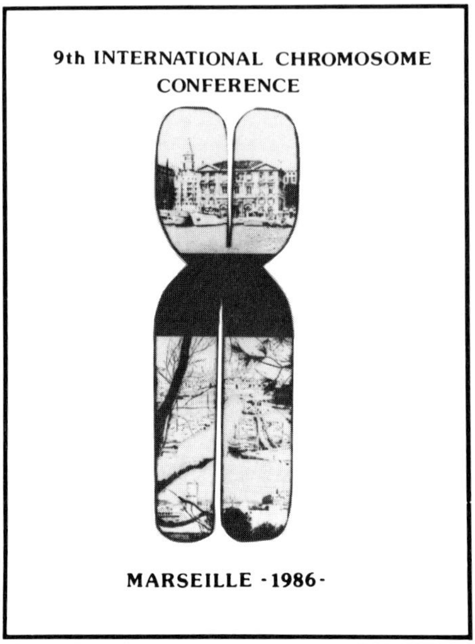

ORGANISED BY
Professor A. Boué; Dr J. Boué, Dr B. Dutrillaux
Dr J. de Grouchy; Professor J.M. Luciani
Professor A. Stahl; Professor A. M. Vagner-Capodano
Professor H. van den Berghe

This was the ninth of the International Chromosome Conferences inaugurated by C. D. Darlington and K. R. Lewis, held in Oxford in 1964, 1967 and 1970, in Jerusalem in 1972, in Leiden in 1974, in Helsinki in 1977, in Oxford in 1980 and in Lübeck in 1983. The proceedings of the conferences have appeared under the title Chromosomes today, *volumes 1–8.*

Conference members

Adler I.D. Institut für Säugetiergenetik, GSF, D-8042 Neuherberg, West Germany.
Al Achkar W. Commissariat à l'Energie Atomique, CEN/FAR BP 6, 92265 Fontenay-aux-Roses, France.
Almeida-Toledo L.F. Departamento de Biologia, Instituto de Biociências, Universidade de Sao Paulo, Rua do Matão n° 277 Butanta, SP Cx. postal 11461, CEP 05499,Sao Paulo, Brasil.
Alonso C. Instituto de Biologia Molecular, Facultad de Ciencias, Madrid, Spain.
Andre M.J. Laboratoire de Cytologie et Cytogénétique, Hôtel Dieu, Place Alexis Ricordeau, 44035 Nantes Cedex, France.
Antoine J.L. Ludwig Institute for Cancer Research, 74 Av. Hippocrate (UCL 7459) 1200/Brussels, Belgium.
Augusseau S. Laboratoire de Cytogénétique, Hôpital des Sablons, 38700 La Tronche, France.
Ashley T. Department of Zoology, University of Tennessee, Knoxville Tennessee 37996-0810, USA.
Aulard S. Laboratoire de Biologie et Génétique Evolutives-CNRS-91190 Gif-sur-Yvette, France.
Aurias A. Institut Curie, Section de Biologie, 26, rue d'Ulm, 75231 Paris Cedex 05, France.
Autio K. Department of Medical Genetics, University of Helsinki, Helsinki, Finland.
Avivi Y. Department of Plant Genetics, Weizmann Institute of Sciences, Rehovot 76100, Israël.
Avivi L. Department of Human Genetics, Sackler Faculty of Medicine, Tel Aviv University, Tel Aviv, Israël.
Ayme S. Centre de Génétique Médicale, Hôpital d'Enfants de la Timone, Bd J. Moulin-Rue St Pierre, 13385 Marseille Cedex 5, France.
Ayraud N. Centre de Génétique Médicale, Faculté de Médecine de Nice, Chemin de Valombrose, Nice 06, France.
Bastard C. Centre Régional de Transfusion Sanguine, B.P. 58, 76232 Bois Guillaume Cedex, France.
Batanian J. Regional Genetics Centre, East Birmingham Hospital, Bordesley Green East, Birmingham B9 5ST, UK.
Beermann F. Institut für Humangenetik der Universität Göttingen, Gosslerstrasse 12 d D-3400 Göttingen, West Germany.
Bellec V. Laboratoire de Diagnostic Anténatal-Cytogénétique, Centre Hospitalier Régional, 44035 Nantes Cedex, France.
Benet Catala J. Departamento de Biologia Celular y Fisiologia, Universidad Autonoma de Barcelona, Barcelona, Spain.
Bennett M. Plant Breeding Institute, Maris Lane, Trumpington, Cambridge CB2 2LQ, UK.
Berger R. Laboratoire de Cytogénétique, Centre Hayem, Institut de Recherches sur les Maladies du Sang, Hôpital St Louis, 2, Place du Docteur Fournier, 75475 Paris Cedex 10, France.
Bernard Ph. Laboratoire d'Hématologie, Hôpital Cardiologique Haut-Leveque, Avenue Magellan, 33604 Pessac, France.
Beverstock G. Department of Human Genetics, Sylvius Laboratory, Wassanaazseweg 72, 2333 Al Leiden, The Netherlands.
Biddle F. Alberta Children's Hospital Research Centre, 1820 Richmond Road S.W. Calgary, Alberta, Canada T2T 5C7.
Bignon Y.J. Centre Jean Perrin B.P. 392, Place H. Dunant, 63011 Clermont Ferrand Cedex, France.
Bornand-Meglioli G. Rue du Casino 33, Montreux, Suisse.

Boue A. Centre International de l'Enfance, Château de Longchamp, Bois de Boulogne, 75016 Paris, France.
Boue J. Centre International de l'Enfance, Château de Longchamp, Bois de Boulogne, 75016 Paris, France.
Bourgeois C. Laboratoire de Pathologie Cellulaire, 15, rue de l'Ecole de Médecine, 75270 Paris Cedex 06, France.
Bova R. Istituto di Genetica Umana, Universita Cattolica, Largo Francesco Vito 1, 00168 Roma, Italy.
Brahe C. Istituto di Genetica Umana, Universita Cattolica, Largo Francesco Vito 1, 00168 Roma, Italy.
Breed A. Antonius Deusinglaan 4, 9713 AW Groningen, The Netherlands.
Bresson J.L. Laboratoire de Cytogénétique, Centre Hospitalier et Universitaire St Jacques, 25000 Besançon, France.
Britton-Davidian J. Laboratoire de Génétique, Institut des Sciences de l'Evolution, USTL, Place E. Bataillon, 34060 Montpellier Cedex, France.
Brüderlein S. Institut für Humangenetik, Schwabachanlage 10, D-8520 Erlangen, West Germany.
Bühler E.M. Department of Genetics, Kinderspital, Römergasse 8, CH-4005 Basel, Switzerland.
Bussani C. Human Genetics Service, Department of Paediatries, Via L. Giordano 13, University of Florence, 50132, Italy.
Buvat R. Laboratoire de Génétique, Faculté de Médecine, 27, Bd J. Moulin, 13385 Marseille Cedex 5, France.
Capanna E. Department of Animal and Human Biology, Via Borelli 50, I-00161 Roma, Italy.
de Capoa A. Department of Genetics and Molecular Biology, University of Rome, Via Lancisi 29, Roma 00161, Italy.
Carlon N. Laboratoire de Génétique, Faculté de Médecine, 27, Bd J. Moulin, 13385 Marseille Cedex 5, France.
Chandley A.C. MRC Clinical and Population Cytogenetics Unit, Western General Hospital, Edinburgh EH4 2XU, UK.
de la Chapelle A. Department of Medical Genetics, University of Helsinki, Haartmaninkatu 3, 00290 Helsinki 29, Finland.
Charrin C. Laboratoire Central d'Hématologie-Cytogénétique, Hôpital E. Herriot, 3 Place d'Arsonval, 69374 Lyon Cedex, France.
Chudoba I. Institut für Humangenetik, Bahnhofstrasse 7A, 3550 Marburg, West Germany.
Corti M. Department of Animal and Human Biology, University of Rome, Italy.
Costas E. Hospital "Teresa Herrera", Seccion Genética, La Coruna, Spain.
Couderc H. Laboratoire de Biologie Végétale C, Université Paris Sud, Centre d'Orsay, 91405 Orsay, France.
Couturier J. Institut Curie, Section de Biologie,CNRS, 26, rue d'Ulm, 75231 Paris Cedex 05, France.
Couturier M.H. Laboratoire d'Histologie-Embryologie Cytogénétique, Hôpital Bichat, 46, rue H. Huchard, 75018 Paris, France.
Creau-Golberg N. Laboratoire de Cytogénétique Humaine et Comparée, Hôpital Necker-Enfants Malades, 149, rue de Sèvres, 75743 Paris Cedex 15, France.
Dastugue N. Centre Régional de Transfusion Sanguine, Centre Hospitalier et Universitaire Purpan, 31052 Toulouse Cedex, France.

Delozier-Blanchet C. Institut de Génétique Médicale, CMU, 9, A. de Champel, CH-1211 Genève 4, Suisse.
Derenzini M. Istituto di Patologia Generale, Via S. Giacomo 14, 40126 Bologna, Italy.
Dettmers R. Institute of Human Genetics, Academic Medical Centre, Meibergdreef 15, M. 1105 AZ, Amsterdam, The Netherlands.
Devictor M. Laboratoire de Génétique, Faculté de Médecine, 27, Bd J. Moulin, 13385 Marseille Cedex 5, France.
Dietrich A. Institute of Human Genetics, University of Amsterdam, Academic Medical Centre, Meibergdreef 15, 1105 AZ Amsterdam, The Netherlands.
Dolfini S. Dipartamento di Genetica, Via Celoria 26, 20133 Milano, Italy.
Dutrillaux B. Institut Curie, Section de Biologie, CNRS U.A. 620, 26, rue d'Ulm, 75231 Paris Cedex 05, France.
Edlinber H. Institut für Humangenetik, Universität des Saarlandes, 6650 Homburg, West Germany.
Eichenlaub-Ritter U. Fakultät für Biologie, Universität Bielefeld, Postfach 8640, 4800 Bielefeld 1, West Germany.
Eiben B. Institut für Humangenetik, Universität Göttingen, 34 Göttingen, West Germany.
Feldman M. Department of Plant Genetics, Weizmann Institute of Sciences, Rehovot 76100, Israël.
Fernandez-Piqueras J. Departamento de Genetica, Facultad de Ciencias, Universidad Autonoma de Madrid, Spain.
Ferraro M. Dipartimento di Genetica e Biologia Molecolare, Università "La Sapienza" 00185 Roma, Italy.
Feruci L. Dipartimento di Biologia, Università degli Studi di Roma, Via Orazio Raimondo 00173 (La Romanina) Roma, Italy.
Fisher E. Biochemistry Department, St Mary's Hospital Medical School, London W2 1PG, UK.
Fonatsch C. Institut für Humangenetik, Medizinische Universität zu Lübeck, Ratzeburger Allee 160, D-2400 Lübeck, West Germany.
Ford C.E. 156, Oxford Road, Abingdon, Oxon OX14 2AF, UK.
Fournier E. Rhône-Poulenc Santé, Centre de Recherches de Vitry, 13, quai J. Guesde B.P. 14, 94403 Vitry sur Seine Cedex, France.
Fraccaro M. Istituto di Biologia Generale, Università di Pavia, Via Forlanini 14, 27500 Pavia, Italy.
Fraisse J. Laboratoire de Cytogénétique, Centre Hospitalier Régional Universitaire, 25, Bd Pasteur, 42023 St Etienne Cedex, France.
Fredga K. Department of Genetics, University of Uppsala, Box 7003 S-750 07 Uppsala, Sweden.
Frelat G. Commissariat à l'Energie Atomique, IPSN, B.P.6 92265 Fontenay-aux-Roses, France.
Friebe B. Institut für Pflanzenbau und Pflanzenzüchtung, 8050 Freising-Weihenstephan, West Germany.
Fuster Marques C. Departamento de Biologia Celular y Fisiologia, Universidad Autonoma de Barcelona, Barcelona, Spain.
Garagna S. Dipartimento di Biologia Animale, Piazza Botta 10, 27100 Pavia, Italy.
Gamperl R. Institut für Medizinische Biologie und Humangenetik, Karl-Franzens-Universität, Harrachgasse 21, A-8010 Graz, Austria.

Garcia-Caldés M. Departamento de Biologia Celular y Fisiologia, Universidad Autonoma de Barcelona, Barcelona, Spain.
Gebhart E. Institute of Human Genetics, University of Erlangen-Nürnberg, Schwabachanlage 10, D-8520 Erlangen, West Germany.
Geneix A. Laboratoire d'Histologie Embryologie Cytogénétique, Faculté de Médecine, B.P. 38, 63001 Clermont Ferrand Cedex, France.
Germain D. Laboratoire Central d'Hématologie et de Cytogénétique, Hôpital E. Herriot, 69374 Lyon Cedex, France.
Gilgenkrantz S. Centre Régional de Transfusion Sanguine, Avenue de Bourgogne, 54111 Vandoeuvre, France.
Gimelli G. Via Guardastelle 25/15, 16035 Rapallo (G.E.), Italy.
Giovannucci Uzielli M.L. Human Genetics Service, Department of Paediatrics, University of Florence, Via L. Giordano 13, Florence 50132, Italy.
Giraud F. Centre de Génétique Médicale, Hôpital d'Enfants de la Timone, Bd J.Moulin-Rue St Pierre, 13385 Marseille Cedex 5, France.
Glamann J. Department of Physiology, Carlsberg Laboratory, 10 Gl. Carlsberg vej, DK-2500 Copenhagen Valby, Denmark.
Godward M. School of Biology, Queen Mary College, London E 14NS, UK.
Goessens G. Laboratoire de Cytologie et Histologie Animale, Institut A. Swaen, Faculté des Sciences, Université de Liege, Rue de Pitteurs,20, B-4020 Liege, Belgique.
Gonzales-Aguilera J.J. Departamento de Genética,C-XV, Division de Biologia, Facultad de Ciencias, Universidad Autonoma de Madrid, Cantoblanco, 28049-Madrid, Spain.
Gosalvez Berenguer J. Departamento de Genética, Facultad de Ciencias, Universidad Autonoma, 28043-Madrid, Spain.
Gouget P. Laboratoire de Cytologie Clinique et Cytogénétique, Faculté de Médecine, Avenue Kennedy, 30000 Nîmes, France.
Gouzien M. Laboratoire de Génétique, Faculté de Médecine, 27, Bd. J. Moulin, 13385 Marseille Cedex 5, France.
Goyanes V.J. Section of Genetics, "Teresa Herrera" Hospital, Insalud, La Coruna, Spain.
Grant C.J. Department of Botany, University of Bristol, Bristol, BS8 1UG, UK.
Greilhuber J. Botanisches Institut der Universität Wien, Rennweg 14, A-1030 Wien, Austria.
Gross B.L. Hunter College of the City University of New York, 695 Park Avenue, New York,N.Y.1021, USA.
de Grouchy J. Laboratoire de Cytogénétique Humaine et Comparée, Hôpital Necker-Enfants Malades, 149, rue de Sèvres, 75743 Paris Cedex 15, France.
Grunwald D. Commissariat à l'Energie Atomique-IPSN-DPS/SPE, B.P.6 Fontenay-aux-Roses, France.
Guichaoua M. Laboratoire d'Embryologie et de Cytogénétique, Faculté de Médecine Secteur Nord, Bd B. Dramard, 130015 Marseille Cedex, France.
Guitart M. Clinical Infantil Nen Jesus, C/Bonaigua,36,Sabadell, Spain.
Haapala O. Laboratory of Genetics, Department of Biology, University of Turku, SF-20500 Turku, Finland.

Hagemeijer A. Department of Cell Biology and Genetics, Faculteit der Geneeskunde, Erasmus Universiteit Rotterdam, Postbus 1738, 3000 DR Rotterdam, The Netherlands.
Hamers G. Department of Genetics and Cell Biology, Rijksuniversiteit Limburg, P.O. Box 616, 6200 MD Maastricht, Holland.
Hansmann I. Institut für Humangenetik der Universität Göttingen, Gosslerstrasse 12d D-3400 Göttingen, West Germany.
Hartung M. Laboratoire de Génétique, Faculté de Médecine, 27, Bd J. Moulin, 13385 Marseille Cedex 5, France.
Hauschteck-Jungen E. Zoologisches Institut der Universität Zurich, Winterthurerstrasse 190, 8057 Zürich, Switzerland.
Hayman D. Department of Genetics, The University of Adelaide, Box 498,G.P.O., Adelaide, South Australia 5001.
Henderson A. Hunter College, Department of Biological Sciences, 695 Park Avenue, New York, N.Y. 10021, USA.
Hennig W. Genetisch Laboratorium, Faculteit der Wiskunde en Natuurwetenschappen, Toernooiveld, 6525 ED Nijmegen, The Netherlands.
Henn W. Institut für Humangenetik, Universität des Saarlandes, 6650 Homburg, West Germany.
Hens L. Department of Medical Genetics, Vrije Universiteit Brussel, Laarbeeklaan 103, B-1090 Brussel, Belgium.
Hernandez D. Laboratoire de Pathologie Cellulaire, 15, rue de l'Ecole de Médecine, 75270 Paris Cedex 06, France.
Heyting C. Institut of Human Genetics, University of Amsterdam, Meibergdreef 15, 1105 AZ Amsterdam, The Netherlands.
Hoovers J. Institute of Human Genetics, University of Amsterdam, Meibergdreef 15,1105 AZ Amsterdam, The Netherlands.
Hors-Cayla M.C. INSERM U12, Hôpital des Enfants Malades, 149, rue de Sèvres, 75743 Paris Cedex 15, France.
Hubert J. Laboratoire de Pathologie Cellulaire, 15, rue de l'Ecole de Médecine, 75270 Paris Cedex 06, France.
Hulten M. Regional Cytogenetics Laboratory, East Birmingham Hospital, Bordesley Green East, Birmingham B9 5ST, UK.
Hummler E. Institut für Humangenetik der Universität Göttingen, Gosslerstrasse 12d D-3400 Göttingen, West Germany.
Hustinx T. Department of Human Genetics, University of Nijmegen, Nijmegen, The Netherlands.
Jenkins G. Department of Agricultural Botany, School of Agricultural Sciences, U.C.W. Penglais, Aberystwyth, Dyfed, SY 23 3 DD, UK.
Jhanwar S.C. Department of Pathology, Memorial Sloan Kettering Cancer Center, 1275 York Avenue, New York, N.Y. 10021,USA.
Johannisson R. Institut für Pathologie, Medizinische Universität zu Lübeck, Ratzeburger Allee 160, D-2400 Lübeck, West Germany.
Jones K.W. Department of Genetics, University of Edinburgh, King's Buildings, West Mains Road, Edinburgh EH9 3JN, UK.
Jones R.N. Department of Agricultural Botany, University College of Wales, Aberystwyth, SY 23 3 DD, UK.
Jones K. Royal Botanic Gardens, Kew, Richmond, Surrey, UK.
Jordan E.G. Department of Biophysics, Cell and Molecular Biology, King's College London, 26-29 Drury Lane, London WC 2B 5RL, UK.
Kaiser P. Institut für Humangenetik und Genetische Poliklinik, Philipps-Universität, Bahnhofstrasse 7A, D-3550 Marburg, West Germany.

Kalweit S. Institute of Genetics, Arnimallee 5-7 FUB, D-1000 Berlin 33, West Germany.
Kenton A. Cytogenetics Section, Jodrell Laboratory, Royal Botanic Gardens, Kew, Richmond, Surrey, UK.
Khalfallah N. Laboratoire de Génétique et Physiologie du Développement des Plantes, CNRS, 91190 Gif-sur-Yvette, France.
Klasterska I. Wallenberg Laboratory, University of Stockholm, 106 91 Stockholm, Sweden.
Kniewald A. Institut für Humangenetik und Anthropologie, Universität Erlangen, Bismarckstrasse 10, D-8520 Erlangen, West Germany.
Knuutila S. Department of Medical Genetics, University of Helsinki, Haartmaninkatu 3,00290 Helsinki, Finland.
Koske-Westphal T. Labor. Dres. Fenner, Bergstrasse 14, D-2 Hamburg 1, West Germany.
Koulischer L. Institut de Morphologie Pathologique, 41, allée des Templiers, 6270 Loverval, Belgium.
Ladenstein R. St. Anna-Kinderspital, 1090 Vienna, Kinderspitalg. 6, Vienna, Austria.
Lafi A. Biomedical and Biophysical Research Laboratory, School of Biological Sciences, Swansea, SA2 8PP,UK.
Laird C. Department of Zoology, University of Washington, Seattle, Washington 98195, USA.
Lamberti L. Departimento di Biologia Animale dell Università di Torino, Via Accademia Albertina, 17, Torino, Italy.
Lasne C. Institut de Recherches Scientifiques sur le Cancer, CNRS, B.P.N°8, 94802 Villejuif Cedex, France.
Leipoldt M. Human Genetics Institute, University of Goettingen, D 3400 Goettingen, West Germany.
Lemeunier F. Laboratoire de Biologie et Génétique Evolutive,CNRS, 91190 Gif-sur-Yvette, France.
Leonard C. Laboratoire d'Histologie Embryologie Cytogénétique, Centre Hospitalier de Bicêtre, 78, rue du Général Leclerc, 94270 Kremlin-Bicêtre, France.
Leporrier N. Laboratoire de Cytogénétique, Centre Hospitalier et Universitaire Côte de Nacre, 14000 Caen, France.
Leroux D. Laboratoire de Cytogénétique, Centre Hospitalier et Universitaire, 38043 Grenoble Cedex, France.
Lessard M. Laboratoire d'Hématologie, Centre Hospitalier B.P.24, 76083 Le Havre Cedex, France.
Levan G. Department of Genetics, University of Göteborg, Po Box 33031 S40033 Göteborg, Sweden.
Linnert G. Institut für Angewandte Genetik, Freie Universität Berlin (WE6)(FB23), Albrecht-Thaer-Weg6, D-1000 Berlin 33, West Germany.
Longy M. 31, rue de Soissons, Bordeaux 33000, France.
Lopez M.C. Instituto de Biologia Molecular, Facultad de Ciencias, Universidad Autonoma de Madrid, Canto Blanco, Madrid, Spain.
Lovering R. Department of Genetics, The University, Sheffield S10 2IN, UK.
Luciani J.M. Laboratoire d'Embryologie et de Cytogénétique, Faculté de Médecine Secteur Nord, Bd B. Dramard, 13015 Marseille Cedex, France.
Mace B. 24, rue du Faubourg St Jacques, 75674 Paris 14, France.
Madan K. Institute of Human Genetics, Free University of Amsterdam, Meibergdreef 15, 1105 AZ Amsterdam, The Netherlands.

Magenis E. Cytogenetics Laboratory, Oregon Health Sciences University, CDRC 2279, 707 S.W. Gaines Road, Portland, Oregon 97201, USA.

Malet P. Laboratoire d'Histologie Embryologie Cytogénétique, Faculté de Médecine B.P. 38, 63001 Clermont Ferrand Cedex, France.

Mandahl N. Department of Clinical Genetics, University Hospital S-22185 Lund, Sweden.

Mannens M. Institute of Human Genetics, University of Amsterdam, Meibergdreef 15, 1105 AZ Amsterdam, The Netherlands.

Marchi A. Istituto di Biologia Generale, Universita di Cagliari, Via Ospedale 119, 09100 Cagliari, Italy.

Marlhens F. Institut Curie, UA 620 CNRS, 26, rue d'Ulm, 75005 Paris, France.

Mattei J.F. Centre de Génétique Médicale, Hôpital d'Enfants de la Timone, Bd J. Moulin-Rue St Pierre, 13385 Marseille Cedex 5, France.

Mattei M.G. Centre de Génétique Médicale, Hôpital d'Enfants de la Timone, Bd J. Moulin-Rue St Pierre, 13385 Marseille Cedex 5, France.

del Mazo J. Departamento de Genetica, Centro de Investigaciones Biologicas (C.S.I.C.) Velasquez, 144, 28006 Madrid, Spain.

Melcion C. Rhône-Poulenc Santé, Centre de Recherches de Vitry, 13 quai J. Guesde, B.P. 14, 94403 Vitry sur Seine Cedex, France.

Mezzanotte R. Istituto di Biologia Generale, Via Ospedale 119, Cagliari 0900, Italy.

Mikkelsen M. Department of Medical Genetics, John F. Kennedy Institute, Gl. Landevej 7-9, DK-2600 Glostrup, Denmark.

Miller A. Laboratoire de Biochimie Moléculaire, Faculté de Médecine, Université de l'Etat, 24, Av. du Champ de Mars, 7000 Mons, Belgique.

Miller K. Abteilung Humangenetik der Medizinische Hochschule, Postfach 610180 D-3000 Hannover 61, West Germany.

Moirot H. Laboratoire d'Anatomie Pathologique, Hôpital Ch. Nicolle, 1, rue de Germont, 76038 Rouen Cedex, France.

Moraine C. Unité de Génétique, Centre Hospitalier et Universitaire Bretonneau, 2, Bd Tonnele, Tours, France.

Moreau N. Laboratoire d'Histologie, Faculté de Médecine, 8, Avenue Rockefeller, 69008 Lyon, France.

Morescalchi A. Istituto di Anatomia Comparata, Università di Genova, Via Balbi 5, 16126 Genova, Italy.

Muleris M. Institut Curie, UA620 CNRS, 26, rue d'Ulm, 75005 Paris, France.

Muratori R. Università di Genova, Via Balbi 5, Genova, Italy.

Nanda I. Institut für Humangenetik, Universität Berlin, Heubnerweg 6, D-1000 Berlin 19, West Germany.

Narayan R. Department of Agricultural Botany, University College of Wales, Penglais, Aberystwyth Difed SY 23 3 DD, Wales, UK.

Navarro J. Departamento de Biologia Celular Y fisiologia, Universidad Autonoma de Barcelona, Barcelona, Spain.

Novello F. Istituto di Patologia Generale, Università di Bologna, Via S. Giacomo 14, Bologna 40126, Italy.

Nowak C. Institute of Genetics, Arnimallee 5-7 FUB, D-1000 Berlin 33, West Germany.

Ockey C.H. Paterson Laboratories, Christie Hospital, Withington, Manchester, UK.

Ohno S. Beckman Research Institute of the City of Hope, 1450 East Duarte Road, Duarte, California 91010-0269, USA.
Olin-Fatih M. Department of Crop Genetics and Breeding, Swedish University of Agricultural Sciences, Svalov S-26800, Sweden.
Olmo E. Dipartimento di Biologia Evolutiva e Comparata, Università di Napoli, Via Mezzocannone 8 80134 Napoli, Italy.
Pages M.P. Laboratoire de Biologie Médicale, Hôpital E. Herriot, 69374 Lyon Cedex 08, France.
Papes D. Department of Botany, Faculty of Sciences, University of Zagreb, POB 933, Yugoslavia.
Parry E.M. Department of Genetics, University College of Swansea, Singleton Park, Swansea SA2 8PP, UK.
Pasquali F. Istituto di Biologia Generale e Genetica Medica, Università di Pavia, CP 217 27100 Pavia, Italy.
Passarge E. Institut für Humangenetik, Universitätsklinikum der Gesamthochschule Essen, Hufelandstrasse 55, D-4300 Essen 1, West Germany.
Pawlowitzki I.H. Department of Medical Cytogenetics, Institute for Human Genetics, Westfälische Wilhems-Universität, D-4400 Münster, West Germany.
Pearson P. Department of Human Genetics, Sylvius Laboratoria, Instituut voor Anthropogenetica, Waasenaarseweg 72, 2333 AL Leiden, The Netherlands.
Pelliccia F. Dipartamento di Genetica e Biologia Molecolare, Università "La Sapienza", Roma, Italy.
Pession A. Istituto di Patologia Generale, Università di Bologna, Via S. Giacomo 14, 40126 Bologna, Italy.
Pestarino M. Istituto di Anatomia Comparata, Università di Genova, Via Balbi 5, 16126 Genova, Italy.
Petersen M.B. Department of Medical Genetics, John F. Kennedy Institute, 7-9 Gl. Landevej, DK-2600 Glostrup, Denmark.
Petitpierre E. Departamento de Biologia, Facultad de Ciencias, 07071 Palma de Mallorca, Spain.
Pfeiffer R. Institut für Humangenetik, Bismarckstrasse 10, 8520 Erlangen, West Germany.
Polani P. Paediatric Research Unit, Guy's Hospital Medical School, University of London, Guy's Tower, London Bridge, London SE1 9RT, UK.
Prantera G. Dipartimento di Genetica e Biologia Molecolare, Università "La Sapienza", Roma, Italy.
Probeck H.D. Institut für Humangenetik, Bahnhofstrasse 7A, D3550 Marburg, West Germany.
Quack B. Centre Départemental de Transfusion Sanguine, 73011 Chambery, France.
Raimondi E. Via S. Epifanio 14, Pavia 27100, Italy.
Ramella L. Dipartimento di Biologia Animale, Università di Torino, Via Accademia Albertina 17, Torino, Italy.
Raynaud S. Laboratoire de Cytogénétique, Faculté de Médecine de Nice, Chemin de Valombrose, Nice 06, France.
Redi C.A. Dipartimento de Biologia Animale, Università di Pavia, Piazza Botta 10, 27100 Pavia, Italy.
Reeves B. Section of Human Genetics, Institute of Cancer Research, Royal Marsden Hospital, Fulham Road, London SW 3655, UK.
Requena J.M. Instituto de Biologia Molecular, Facultad de Ciencias, Madrid, Spain.
Riviere D. Laboratoire de Cytogénétique, Centre Hospitalier et Universitaire Morvan, 29200 Brest, France.

Robert-Nicoud M. Max-Planck-Institut für Biophysikalische Chemie, Abteilung Molekulare Biologie, D-3400 Göttingen-Nikolausberg, West Germany.

Rocchi A. Dipartimento di Genetica e Biologia Molecolare, Università "La Sapienza", 00185 Roma, Italy.

Rosset R. Laboratoire de Génétique et Biologie Cellulaires, Centre National de la Recherche Scientifique, 13288 Marseille Cedex 9, France.

Ruffie M. Laboratoire de Cytogénétique, Hôpital Pellegrin, Place A. Raba Léon, 33076 Bordeaux, France.

Ruiz-Rejon M. Departamento de Genetica, Facultad de Ciencias, Granada-18071, Spain.

Rumpler Y. Institut d'Embryologie, Faculté de Médecine, 11 rue Humann, 67085 Strasbourg Cedex, France.

Saadallah N. Regional Genetics Centre, East Birmingham Hospital Bordesley Green East, Birmingham B9 5ST, UK.

Sabatier L. Commissariat à l'Energie Atomique CEN/FAR B.P.6, 92265 Fontenay-aux-Roses, France.

Sanchez L. Departamento de Genetica, Centro de Investigaciones Biologicas (C.S.I.C.) Velasquez, 144, 28006 Madrid, Spain.

San Juan A.B. Departamento de Genetica, Centro de Investigaciones Biologicas (C.S.I.C.) Velasquez, 144, 28006 Madrid, Spain.

Saura R. Laboratoire de Cytogénétique, Hôpital Pellegrin, Place A. Raba Léon, 33076 Bordeaux Cedex, France.

Scheer U. Institute of Cell and Tumor Biology, German Cancer research Center, D-6900 Heidelberg 1, West Germany.

Schempp W. Institut für Humangenetik und Anthropologie, Albert-Ludwigs-Universität, Albertstrasse 11, 7800 Freiburg i. Br., West Germany.

Schlegel R. Zentralinstitut für Genetik und Kulturpflanzenforschung der Akademie der Wissenschaften der DDR, 4325 Gatersleben, DDR.

Schmickel R.D. Department of Human Genetics, The school of Medicine, University of Pennsylvania, 37th and Hamilton walk, Philadelphia, PA 19104-6072 USA.

Schmidt A. Institut für Humangenetik, Hufelandstrasse 55, D-4300 Essen 1, West Germany.

Schvartzman J. Departamento de Genetica, Centro de Investigaciones Biologicas (C.S.I.C.) Velasquez, 144, 28006 Madrid, Spain.

Schwarzacher H.G. Histologisch-Embryologisches Institut der Universität Wien, Schwarzspanierstrasse 17, A-1090 Wienn, Austria.

Schwarzacher-Robinson T. Experimental Pathology Group, Los Alamos National Laboratory Life Sciences Division, Los Alamos NM 87545, USA.

Schweizer D. Institut für Botanik und Botanischer Garten der Universität Wien, Rennweg 14 A-1030 Wien, Austria.

Searle J. School of Biological Sciences, University of East Anglia, Norwich, NR4 7TJ, UK.

Segarra C. Departamento de Genetica, Facultad de Biologia, Diagonal 645, 08028 Barcelona, Spain.

Sele B. Laboratoire de Cytogénétique et Biologie de la Reproduction, Faculté de Médecine de Grenoble, 38700 La Tronche, France.

Selva J. Laboratoire d'Histologie, Hôpital de Kremlin-Bicêtre, 94270 Le Kremlin-Bicêtre, France.

Sentis Castano C. Departamento de Genetica, Facultad de Ciencias, Università Autonoma de Madrid, Madrid, Spain.
Serra A. Institute of Human Genetics, "A. Gemelli" School of Medicine, Rome, Italy.
Serville F. Unité de Génétique, Hôpital Pellegrin, Place A. Raba Léon, 33°76 Bordeaux Cedex, France.
Siebert E. Institut für Humangenetik, 8520 Erlangen, West Germany.
Siljak-Yakovlev S. Laboratoire de Systématique et Ecologie Végétales, CNRS-UA121, Université Paris Sud, 91405 Orsay Cedex, France.
Slater R. Institute of Human Genetics, University of Amsterdam, Meibergdreef 15, 1105 AZ Amsterdam, The Netherlands.
Smeets D. Department of Human Genetics, Faculty of Medicine, University of Nijmegen, P.O. Box 9101, 6500 HB Nijmegen, The Netherlands.
Sola L. Department of Animal and Human Biology, University of Rome "La Sapienza", Via Borelli 50, I-00161 Roma, Italy.
Somers A. Biomedical and Biophysical Research, School of Biological Sciences, Swansea SA2 8PP, UK.
Sperling K. Institut für Humangenetik, Universität Berlin, Heubnerweg 6, D-1000 Berlin 19, West Germany.
Spierer P. Département de Biologie Moléculaire, Université de Genève Sciences II, 30, quai E. Ansermet, CH-1211 Genève 4, Suisse.
Stahl A. Laboratoire de Génétique, Faculté de Médecine, 27, Bd J. Moulin 13385 Marseille Cedex 5, France.
Steffensen M. Department of Genetics and Development, College of Liberal Arts and Sciences, University of Illinois at Urbana-Champaign, 515 Morrill Hall, 505 South Goodwin Avenue, Urbana, Illinois 61801, USA.
Stehelin D. Unité d'Oncologie Moléculaire, Institut Pasteur, 1, rue Calmette, B.P. 245, 59019 Lille Cedex, France.
de Stephano G.F. Dipartimento di Biologia, Università degli Studi di Roma, Via Orazio Raimondo, 00173 (La Romanina) Roma, Italy.
Stoll C. Institut de Puériculture, Centre Hospitalier et Universitaire, rue de la Porte de l'Hôpital, 67000 Strasbourg, France.
Taillemite J.L. Hôpital St Antoine, 184, Faubourg St Antoine, 75012 Paris, France.
Tamboise A. Laboratoire d'Histologie Embryologie Cytogénétique, UER Santé Médecine Biologie, 74, rue M. Cachin, 93000 Bobigny, France.
Tamboise E. Laboratoire d'Histologie Embryologie Cytogénétique, UER Santé Médecine Biologie, 74, rue M. Cachin, 93000 Bobigny, France.
Tanzer J. Laboratoire Central d'Hématologie et Cytogénétique Hématologique, Centre de Transfusion Sanguine, rue G. le Troubadour, 86021 Poitiers, France.
Tates A.D. Department of Radiation Genetics and Chemical Mutagenesis, State University of Leiden, Sylvius Laboratories, Wassenaarseweg 72, 2300 RA Leiden, The Netherlands.
Teitelboim E. Department of Genetics, The Hebrew University of Jerusalem, Jerusalem 91904, Israël.
Telvi L. Service d'Histologie-Embryologie, Faculté Cochin-Port-Royal, 24, rue du Faubourg St Jacques, 75674 Paris 14, France.

Temperani P. Istologia-Embriologia Generale, Università di Modena, Via Berengario 16, Modena, Italy.
Templado C. Departamento de Biologia, Facultad de Medicina, Universidad Autonoma, Bellaterra, Barcelona, Spain.
Thepot F. Laboratoire de Biologie de la Reproduction, Centre de Gynécologie-Obstétrique, 124, rue C. Desmoulins, 80000 Amiens, France.
Toga-Piquet C. Laboratoire de Génétique, Faculté de Médecine, 27, Bd J. Moulin, 13385 Marseille Cedex 5, France.
Tolksdorf M. Kinderklinik, Klinikum der Christian-Albrechts-Universität zu Kiel, Schwanenweg 20, 2300 Kiel 1, West Germany.
Tranebjaerg L. Department of Medical Genetics, John F. Kennedy Institute, DK-2600 Glostrup, Denmark.
Traut W. Medizinische Universität zu Lübeck, Institut für Biologie, D-2400 Lübeck, West Germany.
Trepte H.H. Zoologisches Institut-Entwicklungsbiologie, Universität Göttingen D-3400 Göttingen, West Germany.
Turleau C. Laboratoire de Cytogénétique Humaine et Comparée, Hôpital Necker-Enfants Malades, 149, rue de Sèvres, 75730 Paris Cedex 15, France.
Ulmer R. Institut für Humangenetik und Anthropologie, Universität Erlangen, Bismarckstrasse 10, D-8520 Erlangen, West Germany.
Vagner-Capodano A.M. Laboratoire de Génétique, Faculté de Médecine, 27, Bd J. Moulin, 13385 Marseille Cedex 5, France.
Van De Klundert F. Department of Human Genetics, University of Nijmegen P.O. Box 9101 6500 HB, Nijmegen, The Netherlands.
Van Den Berghe H. Centre for Human Genetics, U.Z. Gasthuisberg, Horestraat, 49, B-3000 Leuven, Belgium.
Van Hemel J.O. Department of Clinical Genetics, University Hospital, P.B. 1738, 3000 DR Rotterdam, The Netherlands.
Vanni R. Istituto di Biologia Generale, Via Ospedale 119, Cagliari 0999, Italy.
Vejerslev L.O. Department of Medical Genetics, John F. Kennedy Institute, Gl. Landevej 7, DK-2600 Glostrup, Denmark.
Vekemans M. Department of Paediatric Pathology, The Montreal Children's Hospital, Mc Gill University, Montreal, Canada.
Viegas-Pequignot E. Institut Curie, UA 620 CNRS, 26, rue d'Ulm, 75231 Paris Cedex 05, France.
Viegas M.W. Departamento de Botanica, Instituto Superior de Agronomia Tapada da Ajuda, 1399 Lisboa Codex, Portugal.
Vieira M.R. Departamento de Botanica, Universidade de Coimbra, Coimbra, Portugal.
Viguie F. Laboratoire d'Histologie Embryologie et Cytogénétique, Hôtel Dieu de Paris, 2, rue d'Arcole, 75004 Paris, France.
Viinikka Y. Department of Biology, University of Turku, 20500 Turku, Finland.
Vogel R. Max Von Pettenkofer Institut, Bundesgesundheitsamt, P.O. Box 330013, D-1000 Berlin 33, West Germany.
Voiculescu I. Institut für Humangenetik, Schwabachanlage 10, D-8520 Erlangen, West Germany.
Volleth M. Institut für Humangenetik, Schwabachanlage 10, D-8520 Erlangen, West Germany.
Volobouev V. Génétique Ecologique E.P.H.E., Université de Montpellier-2, Place E. Bataillon, 34060 Montpellier, France.
Wall W. Department of Cytogenetics, Queen Elizabeth Hospital, Hackney Road, London E2 8PS, UK.

Wallace B. Department of Genetics, University of Birmingham, Birmingham, B15 2TT, UK.
Wagenvoort M. Stichting voor Plantenveredeling SVP, Droevendaalsesteeg 1, Postbus 117, 6700 ac Wageningen, The Netherlands.
Wahrman J. Department of Genetics, The Hebrew University of Jerusalem, Jerusalem, Israël.
Weber B. Institut für Humangenetik, Albertstrasse 11, 7800 Freiburg i. Br., West Germany.
Weissenbach P. Unité de Recombinaison et Expression Génétique, Institut Pasteur, 28, rue du Dr Roux, 75724 Paris Cedex 15, France.
Westerman M. Department of Genetics, La Trobe University, Bundoora Melbourne, 3083, Australia.
Wilkinson P. Department of Zoology, University of Oxford, South Parks Road, Oxford, UK.
Wollenberg C. Institut für Humangenetik, Universität des Saarlandes, 6650 Homburg/Saar, West Germany.
Wolf K. Institüt für Biologie der Medizinischen Universität zu Lübeck, Ratzeburger Allee 160, D-2400 Lübeck, West Germany.
Wolf U. Institut für Humangenetik und Anthropologie, Albert-Ludwigs-Universität, Albertstrasse 11, 7800 Freiburg i. Br., West Germany.
Yoon J.S. Department of Biological Sciences, Bowling Green State University, Bowling Green, Ohio 43403-0212, USA.
Zacharias H. Zoologisches Institut der Universität, Biologiezentrum, Olshausenstrasse 40, D-2300 Kiel, West Germany.
Zang K. Institut für Humangenetik, Universität des Saarlandes, 6650 Homburg/Saar, West Germany.
Zuffardi O. Biologia Generale e Genetica Medica, Università di Pavia, c.p. 217, I, 27100 Pavia, Italy.

MOLECULAR CYTOGENETICS

Genetics of heterochromatin in *Drosophila* and maize

D. M. Steffensen

Department of Genetics and Development, University of Illinois, Urbana, Illinois 61801, USA

INTRODUCTION

Over the years there has been a continuing interest in heterochromatin. Brown (1966) divided heterochromatin into two types, the facultative and constitutive heterochromatin. Both are late-labeling with ^{3}H-thymidine in the cell cycle. Facultative heterochromatin is a developmental imprint like the late-labeling X chromosomes. In modern terms, constitutive heterochromatin is defined as the highly repeated DNAs usually made up of short repeats of the same or similar base sequences. This paper will cover our present knowledge concerning some of the highly repeated DNAs in maize and *Drosophila*. This new cytogenetic data will deal with their functions in early development.

The cytogenetic aspects of heterochromatin in animals are best known in *Drosophila* (Hannah 1951; Appels and Peacock 1978). Intercalary heterochromatin is an important feature of the genome. Unlike other highly repeated DNA sequences located around the centromere or at chromosome ends, intercalary heterochromatin is bracketed by functional genes. According to Hannah (1951), intercalary heterochromatin corresponds to repeats, high breakability, and ectopic pairing. These sites are recognized as weak points (underreplication) and by being late-replicating (Arcos-Téran 1972; Zhimulev *et al.* 1982). Weak points are involved in ectopic pairing and have a similar DNA content (Barr and Ellison 1972). At least two major developmental mutants, *bithorax* at 89E (Lewis

1978) and *dorsal-bicaudal* at 36C (Steward and Nüsslein-Volhard 1986) are at weak points (intercalary heterochromatin) in *Drosophila*. *Antennapedia* at 84B12 is bracketed by two weak points at 83CD and 84D. Based on data from Goldschmidt *et al.* (1951), Goldschmidt (1949) concluded that heterochromatin acts early in differentiation. With the *Podotera* effect, a homeotic mutant phenotype that has legs where wings belong, the normal Y chromosome and Y^L chromosomes enhance the penetrance of *Podotera* (pod-M(3)-124).

The same sort of increase occurs in the bobbed-like phenotype to be considered. Y chromosomes or Y^L chromosomes in *Drosophila* species hybrids enhance the phenotype. There are little or no effects when the Y is missing or with Y^S chromosomes. Like Goldschmidt, our main point will be that heterochromatin acts early in development.

In plants, *Zea mays* is best understood, especially several genetic phenomena reviewed by Rhoades (1978). Maize has large segments of DNA called knobs, which are intercalary condensed blocks of heterochromatin made up of an 180 bp repeat (Peacock *et al.* 1977; Dennis and Peacock 1984). An extensive compilation of the maize genome involving the location and frequency of knob DNA has been made by McClintock *et al.* (1981). Their distribution and frequency are described for most races of maize in the Americas. In North America, if one proceeds northward from Mexico to Canada, the number of knobs per nucleus decreases to a point at the northern extremities, only knobless or low knob races are found (Brown 1949; McClintock *et al.* 1981). The genetic experiments presented here will demonstrate that knob homozygosity has a major role in determining the rate of floral development. The knobs appear to act as a developmental switch and apparently regulate its bracketing genes as may be the case in *Drosophila* with intercalary heterochromatin.

In another emphasis on development, a detailed genetic study is presented concerning the heterochromatin of the Y chromosome in *Drosophila* using crosses between *D. melanogaster* and *D. simulans*. The Y chromosome must have a major influence in the differentiation of nuclei at blastoderm and later developmental stages. In several ways our new evidence is a continuation of the classic

Survival Effect of X and Y Chromosomes in Reciprocal Crosses Between
D. melanogaster and D. simulans

	Female		Male	
(Parents)	X_mX_m	x	X_sY_s	
(Progeny)	X_mX_s (adults)	↓	X_mY_s (die as embryos)	
	X_sX_s	x	X_mY_m	
	X_sX_m (die in late 3rd instar)	↓	X_sY_m (adults)	
	$\overline{XX}_mY_m$	x	X_sY_s	
	$\overline{XX}_mY_s$ (die in late 3rd instar)	↓	X_sY_m (adults)	
	X_sX_s	x	$\overline{XY}_mB^SY_my^+$	
	$X_s\overline{XY}_m$ (adults)	↓	$X_sB^SY_my^+$ (adults)	
	$X_m(\underline{B})X_m(\underline{B})$	x	$X_s(v)Y_s$	
	$X_m(\underline{B})X_s(v)$ (adults)	↓	$X_m(\underline{B})0$ (putative, adults, highly bb-like, 1/5,000 in male embryos)	
	$\overline{XX}_m/T(1;4)B^S$	x	X_sY_s	*Both males and females have notched-like wings.
	$\overline{XX}_m/T(1;4)B^S/Y_s$ (adults)	↓	$X_s/T(1;4)B^S$ (adults)*	
	$X_m\underline{y}\ \underline{w}\ \underline{f}/\overline{XY}_m\underline{y}\ \underline{w}^a/y^+Y^S\cdot Y^L\underline{kl}$ 1-5;4R	x	$X_s(v)Y_s$	
	$\overline{XY}_m\ \underline{y}\ \underline{w}^a/X_s$ (adults) and	↓	$\overline{XY}_m\underline{y}\ \underline{w}^a/y^+Y^S\cdot Y^L\underline{kl}^{1-5}$;4R/$Y_s$ (adults, 2/100) and	
	$X_m\ \underline{y}\ \underline{w}\ \underline{f}/X_s$ (adults)		$\overline{XY}\ \underline{y}\ \underline{w}^a/Y_s$ (adults, 3/1,000, could be $\overline{XY}$ 0)	

Table 1 Survival and the effects of X and Y chromosomes in reciprocal crosses between D. melanogaster and D. simulans

A bobbed-like phenotype controlled by sequences on the long arm of the Y (Y^L)

In Table 2 adult males were scored for a bobbed-like or abnormal abdomen-like phenotype using four classes or categories. A ring chromosome Y^L derivative (R(Y)L16) obtained from Dr. W. L. Baker had the most dramatic and extreme effect, especially when the flies were grown at 15° or 18°C. For pupae the temperature sensitive period is immediately after pupation during a 24 hour period when the histoblast cells are dividing and differentiating. Y^S combinations and deletions of the rDNA genes have little or no

work of Sturtevant *et al.* (1929). In the latter volume, it was T. H. Morgan who recognized that the Y chromosome had the major effect in the survival of males in *D. melanogaster*/*D. simulans* hybrids.

RESULTS AND DISCUSSION

This symposium article attempts to touch on the highlights of several genetic and molecular phenomena. Firstly, let us consider the role of the Y chromosome of *Drosophila*.

Sturtevant *et al.* (1929) noted that *yellow* (*y*) *D. melanogaster* females mate more readily with *D. simulans* males than do wild-type females. Taking advantage of this fact, we have used attached X females [C(1)DX,*y f bb*] carrying various Y chromosome derivatives. In this way, the special Y chromosomes go to the sons. The notations, X_sY_m, are used to denote the source of the X and the Y with the subscript s for *D. simulans* and m for *D. melanogaster*. Space does not permit the detailed description of the special Y chromosomes, the attached X-Y constitutions, or any other special stocks employed. A detailed account will appear elsewhere. The basic features are presented in Table 1 for survival of the *D. melanogaster-simulans* hybrids. A variety of X and Y combinations are constructed for males and females. In general, male hybrids survive with the X of *D. simulans* (X_s). In fact, X_sO males are the most vigorous of all of the constructed genotypes, indicative of the detrimental role in development for Y_m. We will demonstrate that certain segments of the *D. melanogaster* Y (Y_m) are responsible for the detrimental or imperfect program in development. Occasionally X_mO or attached X_mY_m males will survive but at low frequency. Hybrid females survive best with the X_sX_m combination. Homozygous X_m combinations survive adults if Y heterochromatin is present. In general the life or death genetic responses in these species hybrids is entirely dependent on the presence or absence of "genetically inactive" Y heterochromatin in these species hybrids. The *bobbed*-like phenotype is shown in Figure 1.

Genotype	*simulans* Strain	temp °C	Percentage of Bobbed-Like Phenotype				Total
			bb^+	Light *bb*	*bb*	Extreme *bb*	
X_s/R(Y)L16	Hawaii	18	0	0	61.8	38.1	76
		18	0	0	33.3	66.6	27
		24	0	6.0	77.3	16.5	115
	v	24	0	0	2.4	97.6	42
	Texas	18	0	2.6	86.9	10.4	153
		24	1.4	41.4	47.1	10.0	70
	Honduras	18	0	4.1	37.5	58.3	24
		24	0	6.1	57.9	36.0	114
	Hawaii	24	55.6	38.3	6.0	0	133
X_s/Y $R(1)^L15$	*v*	24	79.1	14.1	6.7	0	163
	Honduras	24	87.7	10.9	1.3	0	73
	Honduras	24	73.1	20.7	6.1	0	82
X_s/y^+y^S#5	Hawaii	24	74.5	11.8	13.6	0	110
	v	18	0	0	75	25	16
		24	0.5	66.3	31.6	1.5	196

Table 2 Influence of different wild type strains of *D. simulans* crossed to *D. melanogaster* with special Y chromosomes in producing a *bobbed*-like phenotype

effect on abdominal development. As stated before, X_sO males are the most wild-type and the most vigorous adults, again indicative of no NOR or ribosome involvement.

Deleted Y^L chromosomes are used in crosses (Table 3). The total deletion of Y^L segments give uses normal adult phenotype and shorter deletion being intermediate. Our cytological examination

Genotype	*D. simulans* Strain	Percent of Bobbed-Like Phenotype				Total
		bb^+	Light *bb*	*bb*	Extreme *bb*	
X_s/y^+Y *kl*-45N	*v*	0	4.0	58.7	37.3	271
X_s/y^+Y *kl*-5Br	*v*	0	9.3	77.3	13.4	97
X_s/y^+Y *kl*-45W	*v*	87.4	7.1	5.5	1.6	127
X_s/y^+Y *kl*-45W	Hawaii	96.0	4.0	0	0	175

*These *kl* stocks were obtained from Dr. T. Kaufman.

Table 3 Additional deletion stocks for male-fertility factors (*) involving the long arm of Y^L (melanogaster) and their influence on male species hybrids at 19°C

of $y^{+}Y$ kl-45W shows it to be essentially deficient for all of the long arm.

Temperature sensitivity of embryo and organ development

The male species hybrid X_s, R(Y)L16, is extremely sensitive to a shift down of temperature (24°C to 15°C) during embryogenesis.

Embryo Age (Hours)	Defect Produced by 15°C in Embryos	Survival Stage
1		adult males
2		
3	thin trachea, missing trachea, missing Malpighian tubes, abnormal trachea	black pupal
4		death at 1st or 2nd instar, variegated bristles on larvae
5	no mouth parts at 1st instar, missing Malpighian tubes	death at early 1st instar
6	thin Malpighian tubes	
7	asymmetric trachea	1st instar death
8	missing right or left fat bodies	
9	beaded fat body	dead 2nd instar
10	no mouth parts or head	
11		
12		
13	thin mouth parts with missing connectors	dead 3rd instar
14		
15		dead pupae, absence of eye development
16		few adult males
17		
18		
19		eyes develop, late pupal death
20		
21		
22		eye development, black pupal death

Table 4 Survival and abnormalities produced by temperature pulse at 15°C to hybrid embryo from cross $\widehat{XX}$ R(Y)L16 x D. simulans (v)

Even an hour or two at 15°C will entirely suppress organ development during blastoderm development and early embryonic stages (Table 4). Preliminary cytological analysis of embryos indicates that the low temperature prevents the proper formation of the chromocenter. When dividing cells stop and begin to differentiate at 15°C, they do not seem to form a normally differentiated nucleus with a single center of heterochromatin at one end. Two or three centers are observed (data not presented).

At the time imaginal eye disks are dividing during the first and second instars, if one shifts to 15°C for one or two days then no eye facets appear in the imago at emergence of the adult. These adult eye cells are an amorphous mass of tissue (data not shown).

A molecular explanation for the genetic responses of Y^L

Two highly repeated DNA sequences make up the major part of Y^L. These sequences are $\frac{\text{AATAT}}{\text{TTATA}}$ and $\frac{\text{AATATAT}}{\text{TTATATA}}_N$ with density in CsCl of 1.672. The other repeated sequence is $\frac{\text{AAGAG}}{\text{TTCTC}}_N$ and $\frac{\text{AAGAGAG}}{\text{TTCTCTC}}_N$ with the density of 1.705. Both were described by Brutlag *et al.* (1978). The cytological map for the 1.672 and 1.705 sequences on the Y chromosome are similar for both *D. melanogaster* and *D. simulans* (Steffensen *et al.* 1981; Peacock *et al.* 1977). Both sequences are present on chromosome 2 and 3 in *D. melanogaster* but missing in *D. simulans*. Especially important is the fact that the 1.672 sequence brackets the centromere in *D. melanogaster*. This sequence homology between chromosomes provides an association site that is jeopardized when missing, as with chromosome 2 and 3 in *D. simulans*. An association hypothesis for bringing related sequences together has been proposed by Steffensen (1985) for *D. melanogaster*. This model depends on DNA binding proteins, like D1, that are present in the egg and embryo. In the case of D1, it is known to bind high A:T DNA (Alfagame *et al.* 1980). Centromeric and segmental associations seem to be a major feature of nuclear differentiation in early development.

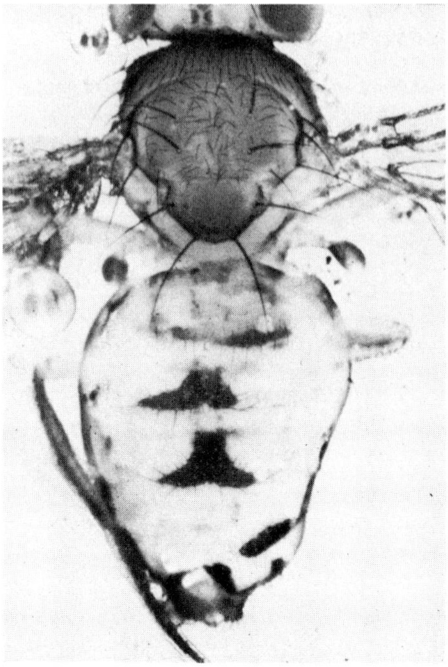

FIGURE 1 A bobbed-like male hybrid between D. melanogaster and D. simulans ($\overline{XX}$ R(Y)L16 x X_s(v)Y_s). The mosaic and major missing bristle areas on the abdomen are produced in this genotype by changing the temperature from 24-25°C to 15°C during the first 24 hours of pupal development. The male in the illustration is classified as an extreme bobbed-like individual. Often these hybrid males (X_sR(Y)L16) are missing segments on one side or the other.

Genetics of knob DNA in maize

Because of space limitations, only a summary for the maize data can be presented. The gist of our data using the chromosomal knobs 9S, 7L, and 4L is that homozygosity for any one or the other of these knobs will extend the time of plant development. A homozygous pair (i.e., 9S K/9S K) is equivalent to a recessive gene delaying development by 3 to 4 days with each pair (i.e., 4L K/4L K and 7L K/7L K) delaying it more. Heterozygotes or homozygous knobless plants are not delayed. Our working hypothesis has these interstitial heterochromatic sequences producing a position effect on their bracketing genes. Perhaps their "late-labeling DNA domain" reduces the amount of gene products.

REFERENCES

Alfagame, R., G. T. Rudkin and L. H. Cohen 1980. Isolation, properties and cellular distribution of D1, a chromosomal protein of Drosophila. Chromosoma 78, 1-31.

Appels, R. and W. J. Peacock 1978.The arrangement and evolution of highly repeated (satellite) DNA sequences with special reference to Drosophila. Intern. Rev. Cytol. Suppl. 8, 69-126.
Arcos-Téran, L. 1972.DNA-replikation und die natur der spät replizurenden orte im X-chromosom von Drosophila melanogaster. Chromosoma 37, 233-296.
Barr, H. J. and J. R. Ellison 1972.Ectopic pairing of chromosome regions containing chemically similar DNA. Chromosoma 39, 53-61.
Brown, S. W. 1966.Heterochromatin. Science 151, 417-425.
Brown, W. L. 1949.Numbers and distribution of chromosome knobs in United States maize. Genetics 34, 524-536.
Brutlag, D., M. Carlson, K. Fry and T. S. Hsieh 1978.DNA sequence organization in Drosophila heterochromatin. Cold Spr. Harb. Symp. Quant. Biol. 42, 1137-1146.
Dennis, E. S. and W. J. Peacock 1984.Knob heterochromatin homology in maize and its relatives. J. Mol. Evol. 20, 341-350.
Goldschmidt, R. 1949.Heterochromatic heredity. Hereditas Suppl. 5, 244-255. Intern. Congress Genetics, Stockholm.
Goldschmidt, R. B., A. Hannah and L. K. Piternick 1951.The Podoptera effect in Drosophila melanogaster. Univ. of Calif. Publ. Zoology 55, 67-294.
Hannah, A. 1951.Localization and function of heterochromatin in Drosophila melanogaster. Adv. Genet. 4, 87-125.
Lewis, E. B. 1978.A gene complex controlling segmentation in Drosophila. Nature 276, 565-570.
McClintock, B., A. Kato Y. and A. Blumenschein 1981.Chromosome constitution of races of maize. Chapingo, Mexico: Colegio de Postgraduados.
Peacock, W. J., A. R. Lohe, W. L. Gerlach, P. Dunsmuir, E. S. Dennis and R. Appels 1981.Fine structure and evolution of DNA in heterochromatin. Cold Spr. Harb. Quant. Biol. 42, 1121-1135.
Rhoades, M. M. 1978.Genetic effects of heterochromatin in maize. In Maize breeding and genetics, D. B. Walden, ed., 641-671. New York: John Wiley.
Steffensen, D. M. 1985.Three-dimensional organization of nuclei of differentiated cells of Drosophila melanogaster. Differentiation 30, 21-28.
Steffensen, D. M., R. Appels and W. J. Peacock 1981.The distribution of two highly repeated DNA sequences within Drosophila melanogaster chromosomes. Chromosoma 82, 525-541.
Steward, R. and C. Nüsslein-Volhard 1986.The genetics of the dorsal-bicaudal D region of Drosophila melanogaster. Genetics 113, 663-678.
Sturtevant, A. H., C. B. Bridges, T. H. Morgan, L. V. Morgan and J. C. Li 1929.Contributions to the genetics of Drosophila simulans and Drosophila melanogaster. Carnegie Inst. of Washington.
Zhimulev, I. F., V. F. Semeshin, V. A. Kulichkov and E. S. Belyaeva 1982.Intercalary heterochromatin in Drosophila. I. Localization and general characteristics. Chromosoma 87, 197-228.

Chromosome accommodation to integration of foreign DNA

A. S. Henderson

Department of Biological Sciences, Hunter College of the City University of New York, 695 Park Avenue, New York, NY 10021, USA

The transfer of foreign DNA into mammalian cells was first described by Szybalska and Szybalski (1962). Subsequent research by many investigators resulted in the development of selection systems for identification of DNA transformed cells. The discovery of covalent linkage between chromosomal DNA and DNA introduced by DNA-mediated transfer led to new approaches for gene mapping and studies of gene regulation in mammalian cells (reviewed in Pellicer et al., 1981). Theoretically, any type of DNA can be stably incorporated into chromosomes, including viral, prokaryotic and non-self (or self) eukaryotic DNA.

Foreign DNA can be introduced into the cell by various means. Experimental methods have included co-precipitation of DNA with calcium phosphate to form a gel (presumably phagocytized by the cell) and microinjection. Viral transformation in vitro, as well as the presence of viral DNA in association with a disease state, has also been shown to result in stable integration of viral DNA. More recently, defective retroviruses have been used as carriers for the introduction of foreign DNA (Hock and Miller, 1986; Williams et al., 1986). Low frequency homologous recombination following DNA transfer has been identified in special cases (Smithes et al., 1986; Thomas et al., 1986), and tissue-specific expression has been described for various genes (Palmiter and Brinster, 1985; Eglitis et al., 1985; Krumlauf et al., 1985; Storb et al., 1985). The successful use of gene transfer methods for human or mammalian gene therapy is inherent in the rationale for many experimental approaches, although the actuality for therapeutic use has remained subject to some debate (Anderson, 1984; Maniatis, 1985; Robertson, 1986).

Over the past five years, we have investigated integration of foreign DNA into mammalian chromosomes from another point of view, that of whether the presence of foreign DNA in the chromosome affects the morphology of the chromosome, and/or increases the propensity for rearrangement. The present report reviews a portion of this research.

RESULTS AND DISCUSSION

Overview. The chromosome complements of cells from a series of experiments where DNA was stably integrated into chromosomes have been analyzed. The use of cloned DNA as probes in molecular hybridizations to both DNA fragments and intact G-banded chromosomes (cytological hybridization) allowed us to correlate alterations in DNA with structural events associated with chromosomes that support integrated DNA. A summary of the experiments used for chromosomal analysis is given in Table 1.

In all cases studied, foreign DNA entered into a single position on one homologue of a chromosome pair. Only one site of stably integrated DNA was present in the genome (Robins *et al.*, 1981a). The integration site was stable with long-term culture of cells, as well as under experimental selection procedures. Cotransformed DNA entered in concert to the same chromosome position. Multiple copies of DNA were usually integrated, but amplification of integrated DNA was also observed (Huberman *et al.*, 1984). The position of the integrated DNA with respect to genomic location was different in independent integrations using the same DNA, which suggests at first level of analysis, that either the integration was random, or, alternatively, that there were many possible positions for integration in the mammalian genome.

A common morphological change in the chromosome supporting integration was the presence of a large achromatic region at the site of integration. The appearance of the achromatic region was consistent with deletion of chromosomal DNA. Modifications of the achromatic regions included constrictions and distortion or deletions in the G-banding pattern. Translocations and other

TABLE 1. Summary of Experiments for Analysis of Chromosomes Following Integration[1]

Source of DNA	Host Cell	Mechanism of Integration	# Integrations Studied	Est. Copy Numbers	Ref.
HGH[3]	rat	CP[2]	4 primary integrations; 5 revertant cell lines from each	HGH:5-100 tk:5-100 other DNA[3]:?	1-3
HSV[2]	mouse	viral transformation	1	1?	4
EBV[2]	human	Burkitt lymphoma	2	1	5,6
	human	transformation *in vitro*	1 + subclone	1	5,6
Xenopus 5S[3]	mouse	CP[2]	2	5S:1-4 tk:1-4 other DNA[3]:?	7
Human beta globin[3]	mouse	microinjection	2	globin:27-68 tk:10-22 pBR:30-90	8
HBV[2]	human	hepatocarcinoma	2	1-2	9
Ad-5[2]	rat	transformation *in vitro*	3	?	unpublished data

[1]The sites of integration of foreign DNA were recognized by cytologic hybridization of cloned ^{125}I-DNA probes (obtained by nick translation using ^{125}I dCTP) to G-banded chromosomes (Henderson, 1982).
[2]Abbreviations: HGH, human growth hormone; CP, calcium phosphate; HSV, *Herpes simplex* virus; EBV, Epstein-Barr virus; HBV, Hepatitis B virus; Ad, adenovirus-5; tk, HSV thymidine kinase gene.
[3]Used HSV tk gene for selection; other DNA = plasmid or carrier.

References: 1. Robins *et al.*, 1981a; 2. Robins *et al.*, 1981b; 3. Henderson and Robins, 1982; 4. Henderson *et al.*, 1981; 5. Henderson *et al.*, 1983; 6. Henderson, 1983; 7. Berg *et al.*, 1983; 8. Huberman *et al.*, 1984; 9. Rogler *et al.*, 1985.

rearrangements that directly involved the integration site were also common. Some specific examples of chromosomal changes involving integration sites are summarized in Table 2.

Evidence for large chromosomal deletions associated with integration. Deletions of cellular DNA in the 10-15 kb range have been previously identified in molecular analyses of integration of DNA-containing viruses (discussed in Rogler *et al.*, 1985). Deletions within this size range would not be detectable at the light microscope level. Yet, visible regions, consistent with the deletion of chromosomal material, were observed at integration sites, irrespective of the nature of the integrated DNA. Other explanations for achromatic regions could be made (see Savage, 1975, for discussion). Our experiments, however, strongly suggested that large deletions were present, and that the deleted endogenous DNA was disproportionately large compared with the size of the integrated DNA.

The first line of evidence was that the dysmorphic chromosome supporting integration was often visibly shorter than the normal homologue; G-bands or portions of G-bands were missing (see Table 2). These abnormalities were accentuated in chromosomes of cells in reversion to the parental phenotype, with accompanying deletion of integrated DNA (Robins *et al.*, 1981b). The second line of evidence was more direct. In one experiment, about 100 copies of human growth hormone DNA integrated at or near a region containing ribosomal DNA (rDNA) on a homologue of rat chromosome 11 (Henderson and Robins, 1982). The effects of foreign DNA insertion could be monitored with respect to endogenous rDNA copy number, as well as for transcriptional activity using NOR-specific silver stain and satellite association frequency as indices of transcription. The outcome of the integration event was inactivation of the rDNA genes at this site, accompanied by the loss of prominent satellites that were present on this chromosome in the original parental cell line. The loss of morphological satellites indicates a large loss of chromosomal material, but in itself, would not necessarily result in a change in the transcriptive activity. Quantitative grain count analysis over chromosome 11 following cytologic hybridization with radioactive ribosomal RNA showed that

TABLE 2. Examples of Chromosomal Changes at Sites of Integration

Abnormality	Source of Foreign DNA	Location of Integrated DNA	Changes Observed
Deletion	HGH DNA and HSV tk DNA	rat 11pter	morphological satellites lost at rDNA site; deletion in rDNA monitored (see text)
		rat 1q (mid-arm)	shorter homologue; G-bands distorted
		rat 4qter	achromatic region (AR); deletion of G-bands
		rat 5qter	AR; deletion of G-bands
	EBV DNA	human 4q25	AR; shorter homologue
		human 1qter	AR; G-band distortion
		human 7q22	AR
	Xenopus 5S	mouse 5ter	affected homologue shorter
	HBV DNA	human 11p	deleted - molecular analysis (Rogler et al., 1985).
		human 17 or 18	as above
	Ad-5	rat Xqter	affected homologue shorter; G-bands deleted
Translocations	HGH DNA	11pter	2 of 5 revertant cell lines with t(3;11) and t(5;11p)
		5qter	revertant cells; multiple translocations
	Xenopus 5S	UK	dicentric chromosome formed at terminal fusion of small unidentifiable chromosomes
	EBV	human 7q	t(6p;7q34) in some Raji cell lines
	HSV	mouse 18 or 19	trisomic chromosome formed
	HBV	human 17 or 18	t(17qter;18q)

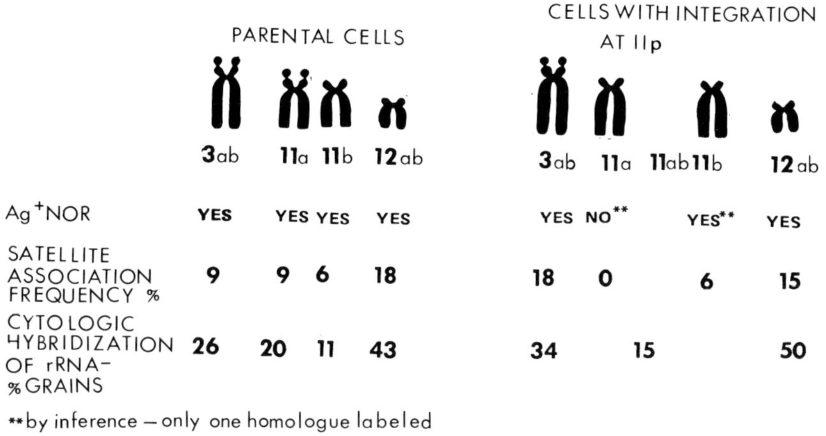

FIGURE 1. Analysis of deletion of rDNA following integration of human growth hormone DNA of rat chromosome 11p. The rDNA sites in the rat are located on chromosomes 3, 11 and 12. a and b indicate homologues; ab: homologues could not be distinguished.

70% of the rDNA previously present at this site in parental cells had been deleted as the result of integration. The amount of rDNA deletion can be calculated as in excess of 1000 kb (illustrated in Figure 1).

What determines integration sites? Recombination between foreign DNA and cellular DNA sharing homology has been established in prokaryotes and yeast, and homologous recombination has been observed between foreign DNA and mammalian chromosomal DNA under special experimental conditions, but at very low frequencies (about 1/1000) (Smithes et al., 1985). In light of the low frequency of homologous recombination at corresponding cellular loci, other predisposing factors that favor integration must be present in the mammalian chromosome. The complexity of the mammalian genome precludes distinguishing between homologous and non-homologous recombination for the majority of integration events. The extensive array of unique and repetitive DNA sequences could direct homologous recombination at many locations on the basis of minimal homology. A second option would favor non-homologous recombination at less specific regions of the chromosome; e.g., regions where integration would not unduly affect chromosome or cellular function. Alternatively, the site chosen for integration could be totally random within the genome.

Theoretically, if homologous recombination plays a role in integration, then foreign DNA could be forced into a site within the recipient chromosome by increasing the number of homologous DNA sequences between the foreign DNA and endogenous cellular DNA. In one series of experiments, we attempted to direct multiple copies of Xenopus 5S DNA to a homologous region in mouse chromosomes containing repetitive 5S DNA (Berg et al., 1983). These experiments were not successful. Although this approach, (with a small sample number), does not provide definitive evidence that forcing integration to a specific site will fail, the results suggested that integration will not be automatically directed to the most prominent regions of DNA homology.

Multiple, partially homologous regions in chromosomal DNA could be responsible for alignment of foreign DNA at many alternative positions in chromosomes. The Epstein-Barr genome has been shown to contain a repeated DNA sequence, IR3 DNA, that has homology to many regions within human (and mouse) chromosomal DNA (Heller et al., 1982). We took advantage of the partial homology to test whether or not small, partially homologous chromosomal regions would facilitate integration, thus mimicking a random distribution. IR3 DNA is internal within the EBV genome and is flanked by unique regions. If EBV DNA recombined with cellular DNA with IR3 homology, then the unique DNA sequences that flank the viral IR3 DNA would be rearranged as the result of recombination. This was not the case, indicating that classical recombination did not occur (Henderson et al., 1983).

The question as to whether a particular chromosome region could be favored for integration, irrespective of homology, remains to be proved. At least two-thirds of the integration sites we have identified, however, were terminal to near terminal. Terminal integration was particularly striking in the chromosomes of cells exposed to thymidine starvation in selection procedures (Robins et al., 1981b). The chromosomes supporting integration were elongated at the terminus initiating at the integration site. Our interpretation of these elongated chromosomes is given in Figure 2. The G-banding pattern suggests that the terminus was deleted,

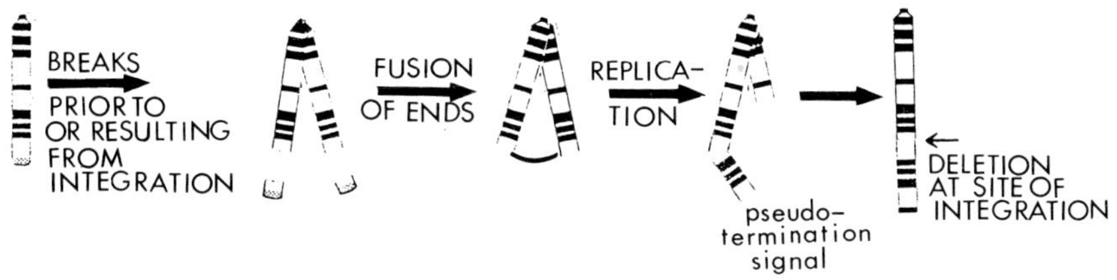

FIGURE 2. One model for the formation of elongated chromosomes at integration sites. Here rat chromosome 5 supported integration of human growth hormone DNA. An elongated chromosome 4 was also observed in an independent integration event.

and that elongation resulted from a reversal of replication to form a longer chromosome. Thus, replication is postulated to have occurred in a continuous fashion around fused termini resulting from breakage until a pseudo-termination site was reached.

Thymidine starvation has been shown to alter normal replication and cause DNA breakage in the chromosome. One explanation, at least for terminal integration sites, is that the transforming DNA takes advantage of ongoing breakage and reunion to stably integrate. Telomeric ends of chromosomes have been shown by many investigators to be susceptible to breakage and rearrangement. Terminal integration sites, as well as achromatic regions, were observed in cells where no experimental selection was involved, e.g., in the integration of viral DNA. Therefore, mechanisms other than the effects of experimental selection could direct integration to terminal sites. The alternative, that transforming DNA generates rearrangements has not been proved, but chromosomal rearrangement has been associated with transpositions (McClintock, 1951).

Many questions remain unanswered, but a major implication of our studies is that the presence of foreign DNA inserted into the chromosome cannot be ignored in terms of chromosome structure. The insertion of large concatenates of DNA from various sources, as well as DNA of DNA-containing viruses, resulted in disproportionate deletions of chromosomal material, as well as a propensity to rearrangement. We hypothesize that aberrant changes will occur in chromosomes as the result of DNA integration, irrespective of the source of foreign DNA, or the method of introduction into the cell. The size of the integrated DNA could determine whether or not structural aberrations would be visible, but the critical experiments have not been done. It should be noted that irrespective of the size of the integration site, chromosome imbalance will occur.

The chromosome as a formed structure exists for the maintenance of the cell. Chromosome sets have accommodated evolutionarily to be species-specific and maintain characteristics that are, at least superficially, unchanging within a species. The general form of the chromosome is retained in spite of retroviral insertion, transposable elements, and structural compensations for individual polymorphic differences. Thus, it is not surprising that the chromosome accepts foreign DNA. It is surprising that aberrant changes could be identified at the visible level. The chromosome will apparently accept self or non-self DNA and compensate by various means for inequities or aberrant differences between homologues.

The integration of foreign DNA into the chromosome occurs and can be associated with abnormalities within the chromosome structure. Yet, perhaps the most important information we have obtained is that the chromosome is an extremely plastic structure. It survives and maintains function in spite of the presence of new DNA and the potential for genic and chromosomal imbalance.

REFERENCES

Anderson, W.F. 1984. Prospects for human gene therapy. Science 226, 401-2.

Berg, P., A. Henderson, S. Ripley, J-K Yu and F. Anderson 1983. Lack of site-specific recombination in mouse L-cells. Biochem. Biophys. Res. Comm. 116, 959-65.

Eglitis, M., P. Kantoff, E. Gilboa, and F. Anderson 1985. Gene expression in mice after high-efficiency retroviral-mediated gene transfer. Science 230, 1395-98.

Heller, M., A. Henderson and E. Kieff 1982. A repeat array in Epstein-Barr virus is related to DNA sequences interspersed on human chromosomes. Proc. Natl. Acad. Sci. USA 79, 5916-20.

Henderson, A., M. Yu and S. Silverstein 1981. Chromosomal DNA homologous to Herpes simplex I virus in a mouse L-cell line. Cytogenet. Cell Genet. 29, 107-15.

Henderson, A. and D. Robins 1982. The effect of exogenous DNA insertion at a chromosomal region containing rDNA. Cytogenet. Cell Genet. 34, 310-14.

Henderson, A. 1982. Cytological hybridization to mammalian chromosomes. Int. Rev. Cytol. 76, 1-46.

Henderson, A., S. Ripley, M. Heller and E. Kieff 1983. Human chromosome association of Epstein-Barr virus DNA in a Burkitt tumor cell line and in lymphocytes growth-transformed in vitro. Proc. Natl. Acad. Sci. USA 80, 1987-91.

Henderson, A. 1983. Localization of viral-specific DNA in the mammalian chromosome complement by cytological hybridization. In Chromosomes and Cancer: From Molecules to Man, J. Rowley and J. Ultman, eds., 247-72. New York: Academic Press.

Hock, R.A. and A.D. Miller 1986. Retrovirus-mediated transfer and expression of drug resistance genes in human haematopoietic progenitor cells. Nature 320, 275-77.

Huberman, M., P. Berg, M. Curcio, J. DiPietro, A. Henderson and F. Anderson 1984. Fate and structure of DNA microinjected into mouse Ltk-cells. Exp. Cell Res. 155, 347-62.

Krumlauf, R., V. Chapman, R. Hammer, R. Brinster and S. Tilghman 1986. Differential expression of α-fetoprotein genes on the inactive X chromosome in extraembryonic and somatic tissues of a transgenic mouse line. Nature 319, 224-26.

Maniatis, T. 1985. Targeting in mammalian cells. Nature 317, 205-6.

McClintock, B. 1951. Chromosome organization and genic expression. Cold Spring Harb. Symp. 16, 13-47.

Palmiter, R. and R. Brinster 1985. Transgenic mice. Cell 41, 343-44.

Pellicer, A., D. Robins, B. Wold, R. Sweet, J. Jackson, I. Lowy, J. Roberts, G-K Sim, S. Silverstein and R. Axel 1980. Altering genotype and phenotype by DNA-mediated transfer. Science 209, 1414-22.

Robertson, M. 1986. Desperate appliances. Nature 320, 213-14.

Robins, D., S. Ripley, A. Henderson and R. Axel 1981a. Transforming DNA integrates into the host cell chromosome. Cell 23, 29-39.

Robins, D., R. Axel and A. Henderson 1981b. Chromosome structure and DNA sequences associated with mutation of transformed genes. Molec. Applied Genet. 1, 93-103.

Rogler, C. M. Sherman, D. Shafritz, J. Summers, T.B. Shows and A. Henderson 1985. Deletion of chromosome 11p associated with Hepatitis B integration site in hepatocellular carcinoma. Science 230, 319-22.

Savage, J. 1975. Classification and relationships of induced chromosomal structural changes. J. Med. Genet. 12, 103-22.

Smithes, O., R. Gregg, S. Boggs, M. Koralewski and R. Kucherlapati 1985. Insertion of DNA sequences into the human chromosomal β-globin locus by homologous recombination. Nature 317, 230-34.

Storb, U., K. Denis, R. Brinster and O. Witte 1985. Pre-B cells in K-transgenic mice. Nature 316, 356-58.

Szybalska, E. and W. Szybalski 1962. Genetics of human cell lines. IV. DNA-mediated heritable transformation of a biochemical trait. Proc. Natl. Acad. Sci. USA 48, 2026-34.

Thomas, K., K. Folger and M. Capecchi 1986. High frequency targeting of genes to specific sites in the mammalian genome. Cell 44, 419-28.

Williams, D., S. Orkin and R. Mulligan 1986. Retrovirus-mediated transfer of human adenosine deaminase sequences into cells in culture and into murine hemapoietic cells in vivo. Proc. Natl. Acad. Sci. USA 83, 2566.

Rapidly evolving Bkm DNA is associated with hypervariable domains

K. W. Jones, E. Olszewska and L. Singh

Department of Genetics, University of Edinburgh, King's Buildings, West Mains Road, Edinburgh EH9 3JN, Scotland

Bkm sequences were first found on the W sex chromosomes of snakes were they exist in amounts large enough to appear as a satellite DNA fraction in females (Singh, Purdom & Jones 1976). In situ hybridization shows that the sequences are interspersed throughout the sex-determining W chromosome (Singh, Purdom & Jones 1980). They are also present in lesser abundance throughout the genomes of all snakes and of all eukaryotes so far surveyed (Jones & Singh 1981a). Concentrations of Bkm sequences occur on sex chromosomes in many other vertebrate species (Jones & Singh 1981b). This ubiquitous distribution does not necessarily indicate functional conservation but could equally reflect processes which generate novel nucleotide sequences. Hybridization of Bkm probes has so far failed to show related sequences in prokaryotes (Jones, Singh & Phillips 1983), suggesting either that the sequences are subject to negative selection or that the chromosomal processes which generate them may be absent.

Bkm satellite DNA includes monotonous arrays of the tetranucleotide GATA (Singh, Phillips & Jones 1984) and the dinucleotides GT and CT (Walker 1985), all of which have been either shown, or suggested to possess the potential for controlling gene expression (Rich, Nordheim & Wang 1984). The ubiquity of poly (dT-dG) in eukaryotes and their absence in prokaryotes has recently been confirmed by Morris, Kushner & Ivarie (1986).

Bkm sequences on sex chromosomes.

The fact that the Z and W chromosomes in the homomorphic ZW bivalents of snake species differ markedly in their Bkm sequence content shows clearly that crossing over was essentially abolished at an early stage in sex chromosome evolution in snakes, and that Bkm sequences increased in its absence. In the homomorphic ZW bivalent, the W chromosome is also heterochromatic in snakes (Singh 1972) and lizards (Olmo, Cobror, Morescalchi & Odierna 1984) which contain many species without specialized sex chromosomes. From this it has been suggested (Jones 1983a) that an early stage, perhaps the essential step, in their evolution involves chromosomal somatic inactivation. This could account both for the inhibition of crossing over and the rapid evolution of the more unstable sequences. It has been proposed that sex chromosome specialization evolved abruptly due to the sex genes assuming control of chromosome condensation centres (Jones 1983a). Such chromosome hijacking mutations rendered the chromosomes (Y and W) bearing the heterogametic sex genes essentially defunct in respect of developmental functions, other than those specified by the sex genes concerned. As a consequence of this abolition of function, selection pressure was removed and sequences with greatest propensity to evolve, such as Bkm GATA repeats, became strikingly abundant in such chromosomes in many species. In contrast, the lower abundance of GATA-rich domains in the autosomes and the presence or absence of polymorphism in individual elements of the GATA blot pattern, as discussed below, reflects selection pressure on these sequences.

In the mouse the genomic DNA blots of all mouse species so far probed with Bkm show a sex pattern difference (our unpublished data). It is clear that the male specific pattern normally originates from a single sex determining segment of Y chromosome DNA (Singh & Jones 1982). Thus, in the sex-reversing Sxr mutation, two such regions are found in the male carrier mouse, one of which becomes meiotically transferred to the X chromosome, giving rise to XX_{Sxr} males whose Bkm blot patterns resemble those of normal males.

Human DNA when blotted and probed with Bkm does not show a sex difference. However, Bkm sequences are relatively abundant in the acrocentric group of chromosomes, including the Y chromosome

(Singh & Jones 1986). The absence of a male-specific blot pattern therefore presumably reflects an interspersed distribution of relatively short Bkm sequences on the Y chromosome.

Bkm sequences and genome evolution.

Bkm sequences are also found throughout the genomes of snakes with or without a specialized W chromosome. However, they are in significantly lesser abundance in snake species lacking a specialized W sex chromosome (Singh, Purdom & Jones 1976). This suggests that their increased abundance is closely connected with the evolution of this chromosome. A suggested mechanism (Jones 1983b) is that such autosomal Bkm sequences originated from the Bkm-rich W chromosome and that the more rapid evolution of species with specialized sex chromosomes has reflected a significant increase in insertional mutations resulting from such interspersion. According to this hypothesis, any evolutionary significance attaching to sex chromosome specialization stems not from their functions in sex determination but from the abrupt cessation of selection pressure on the other 8% or so of genomic DNA carried by these chromosomes. It remains to be established whether GATA-rich domains include significant developmental functions.

Polymorphism in Bkm sequence organization.

In addition to the sex-related patterns in some species, Bkm probed Southern blots show a complex pattern which is individual-specific in all species so far investigated including, for example, in amphibians, turtles (Fig. 1) and humans (Fig. 2). This hypervariability lends itself to genealogical studies in many different species. The individuality of pattern includes any sex-related patterns which may be present. For example, each individual wild male mouse of the same species shows a different Bkm Y chromosome blot pattern whereas that of other repeated sequences, such as mouse satellite, remains unchanged (unpublished). The extent of population polymorphism in GATA-rich RFLPs implies unusually unstable, or hypervariable, chromosomal domains. Evidently gross DNA structure is not conserved in the region of the male determining sex genes. However, elements of the pattern seem to conform to the normal rules of inheritance and the overall pattern is relatively strain-specific in inbred mice. Blot patterns do not alter in normal development.

Sequence variants with a more species-limited distribution but otherwise resembling the present polymorphisms have been described in the vicinity of the globin genes (Jeffreys, Wilson & Thein 1985). These are without apparent affect on the expression of globin genes and are inherited in a Mendelian manner. They are clustered in intervening sequences and are largely absent from the coding sequences or DNA tracts lying between the genes. From the fact that these intervening sequences are not well conserved between species, it appears that any functions this DNA may have does not depend upon a rigidly fixed sequence. Duplication of intervening sequences has also been seen in globin genes and a mechanism of intrachromosomal recombination has been suggested as one cause. It has been suggested by Jeffreys on the basis of globin gene studies that on average 1 in 100 bp could vary polymorphically in man. These might cluster into highly variable hot spots.

In humans, which have been most extensively examined, between 15 and 18 Bkm positive AluI and 17-20 BstNl RFLPs are clearly evident upon probing blots with Bkm GATA sequences (Fig. 2). None of the fragments has so far been mapped to particular chromosomes. Certain of these Bkm-positive restriction fragments are relatively frequent or constant in the populations so far studied, for example the fragments of around 5kb length in Fig. 2, others are less common. Some RFs were more common in individuals sampled from particular populations. For example, in six Australian aborigines, certain fragments were apparent which were absent or very faintly discernable, in blots from DNAs of the same number of caucasians (Singh & Jones 1986). Relative amplification and diminuition in signal strength of bands of particular mobilities also occur in these unrelated individuals. Bands halving or doubling in signal strength mainly reflect heterozygosity and homozygosity for individual fragments and our unpublished preliminary study of the transmission of RFLPs in a three generation family are consistent with this explanation. However, unexplained variations occur which might, for example, involve changes in the relative numbers of GATA repeats per restriction fragment, or varying abundance of tandem repeats of the fragment itself, or to co-migrating different restriction fragments. These possible explanations are being examined.

<u>Somatic polymorphism of Bkm sequences in tumours.</u>

The likelyhood that hypervariable Bkm RFLPs originate from several chromosomes offers the possibility that they may be used to monitor for chromosomal changes in oncogenesis. In a preliminary study (Olszewska, Hay & Jones, submitted) the possible value of hypervariable Bkm RFLPs in elucidating the oncogenic pathways in specific tumours has been investigated. The approach used was to compare normal (leukocyte) DNA with tumour cell DNA from the same patient. In a small cohort of patients with either breast or bladder primary tumours differences relative to normal were noted in the Bkm RFLPs pattern in breast but not in bladder cancers. 8 out of the 10 breast tumour DNAs examined were altered to varying degrees when compared to the respective normal blood DNAs. Three types of intra-individual Bkm RFLP changes were noted. These include loss and gain of bands and relative changes in the hybridization strength of particular bands. Variations were not associated with change in the overall Bkm RFLP pattern. The finding of altered RFLP patterns in breast but not bladder cancers and their limited nature suggests that they may not reflect the common karyotypic pathology of tumours but, hopefully, may be specifically related to the underlying specific causes of some breast cancers. Significantly, in one case, the loss of one band corellated with doubling of the intensity of a separate band. This could be consistent with loss and duplication of specific chromosomes, or chromosomal loci, as found in retinoblastoma and Wilm's tumour. Experiments are in progress to test this working hypothesis.

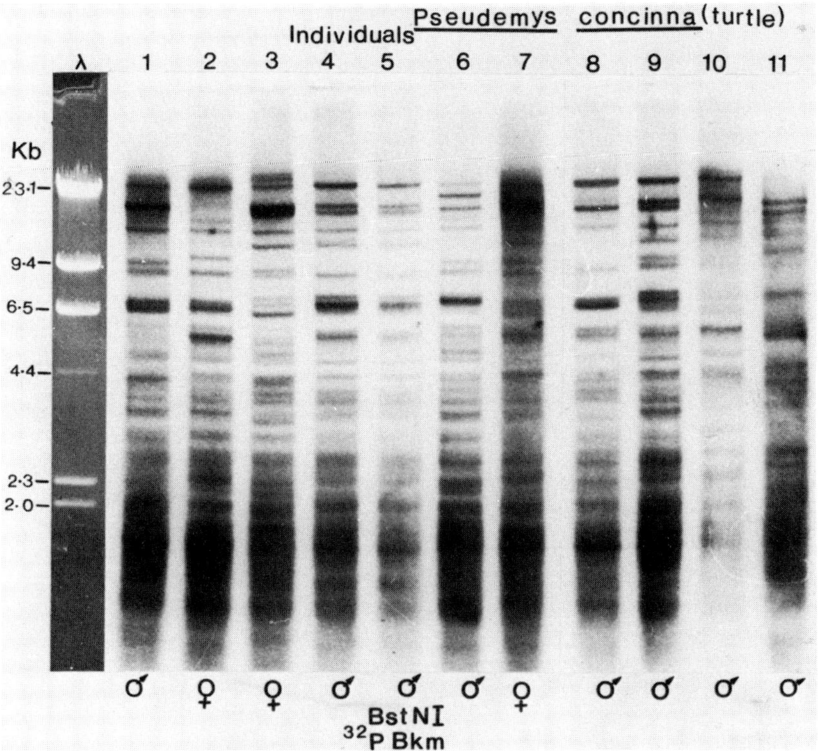

Figure 1. BstN1 digested DNA from individual male and female turtles after Southern transfer and probing with ^{32}P labeled Bkm DNA showing individual polymorphisms. No two individuals are alike and no sex related patterns are discernable.

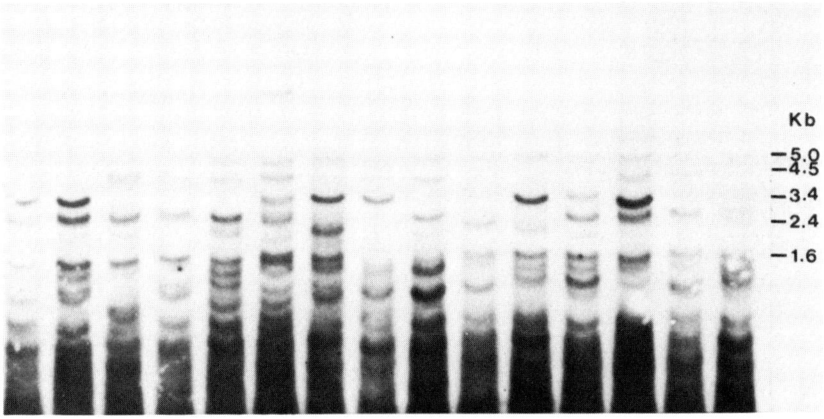

Figure 2. 15 individual Human placental DNAs digested with AluI and probed with Bkm subclone 2(8) (Singh et al 1984) which consists extensively of the sequence $GATA_n$. No two individuals yield the same pattern but certain fragments are consistently present in all individuals.

ACKNOWLEDGEMENTS.

This work was supported bu the Medical Research Council and the Cancer Research Campaign. We thank Mr. K. Chetty and Dr. T. Busuttil for arranging human tumour and blood samples and to Dr. J. Bukl for samples or turtle blood.

REFERENCES.

Jeffries,A.J., Wilson,V., &Thein,S.L. (1985). Hypervariable "mini-satellite" regions in human DNA. Nature 314, 67-73.

Jones,K.W. (1983a). Evolution of sex chromosomes. Development in Mammals 5: 297-230. Ed. Johnson,M.H., Elsevier Science Publishers B.V.

Jones,K.W. (1983b). The evolution of sex chromosomes and their consequences for the evolutionary process. Chromosomes Today 8. (Eds. Bennett,M.D., Gropp,A. and Wolf,U.) Wiley.

Jones,K.W. & Singh,L. (1981a). Conserved sex-associated repeated DNA in vertebrates. In Genome Evolution. G.Dover & R.Flavell, Eds.(London, Academic Press). pp135-154.

Jones,K.W. & Singh,L. (1981b). Conserved repeated DNA sequences in vertebrate sex chromosomes. Hum. Genet. 58, 46-53.

Jones,K.W., Singh,L., & Phillips,C. (1983). Conserved nucleotide sequences on sex chromosomes. Proc. Vth. John Innes Symposium : Genetic Rearrangement. Eds K.F. Chater, C.A. Cullis., D.A. Hopwood, A.A.W.B. Johnston, & H.W. Woolhouse. (Croom Helm London & Canberra) pp265-287.

Morris,J., Kushner,S.R. & Ivarie,R. (1986). Mol. Biol. Evol. 3 (4) 343-355.

Olmo,E., Cobror,O., Morescalchi,A., & Odierna,G. (1984). Homomorphic sex chromosomes in the lacertid lizard Takydromus sexlineatus. Heredity 53 : 457-459.

Rich,A., Nordheim,A., & Wang,A.H-J. (1984). The chemistry and biology of left handed Z DNA. Ann. Rev. Biochem. 53 : 791-846.

Singh,L. (1972) Evolution of karyotypes in snakes. Chromosoma 38: 183-236.

Singh,L., & Jones,K.W. (1982). Sex reversal in the mouse (Mus musculus) is caused bu a recurrent non-reciprocal crossover involving the X and an aberrant Y chromosome. Cell 28, 205-216.

Singh,L. & Jones,K.W. (1986) Bkm sequences are polymorphic in humans and clustered in pericentric regions of various acrocentric chromosomes including the Y. Hum. Genet. 73:304-308.

Singh.L., Purdom,I.F., & Jones,K.W. (1976) Satellite DNA and the evolution of sex chromosomes. Chromosoma (Berl.) 59: 43-62).

Singh,L., Purdom,I.F., & Jones,K.W. (1980). Sex chromosome-associated satellite DNA: Evolution and conservation. Chromosoma (Berl.) 79,137-157.

Singh.1., Purdom,I.F., & Jones,K.W. (1981). Conserved sex chromosome-associated nucleotide sequences in eukaryotes. Cold Spring Harb. Symp. Quant. Biol. 45, 805-813.

Singh,L., Phillips,C., & Jones,K.W. (1984). The conserved nucleotide sequences of Bkm which define Sxr in the mouse are transcribed. Cell. 36, 111-120.

Walker,M.J. (1985). Ph. D Thesis, Edinburgh University.

CHROMOSOMES OF *DROSOPHILA*

Molecular dissection of the *Drosophila* chromosome

P. Spierer

Department of Molecular Biology, University of Geneva, 30 quai Ansermet, 1211 Geneva 4, Switzerland

We have developed methods for gene isolation which do not depend on previous knowledge of the product of the gene (Bender et al. 1983). These protocols, which we called "chromosomal walking and jumping" are now routinely used in most molecular genetics laboratories.

We have isolated two regions of the Drosophila melanogaster chromosome with these methods (Bender et al. 1983). One comprises a group of genes controlling segmentation, the bithorax complex, the other is 315'000 base pairs long and encompasses fifteen well characterized genes including the loci Ace (encoding acetylcholinesterase), rosy (xanthine dehydrogenase), snake (a maternal-effect gene controlling pattern formation polarity), and a suppressor of position-effect variegation (the last two were not known a that time) (Spierer et al. 1983, Gausz et al. 1986).

We have started a detailed structural and genetic analysis of this 315'000 base pairs segment of the Drosophila chromosome. The originality of this approach is to avoid to focus on one particular gene while neglecting its environs. We are trying, on the contrary, to obtain a complete description of an interval containing many functionally unrelated genes. Our aims are to examine the distribution of the functional units on the chromosome, whether essential or not, and to define the extent of essential and non-essential DNA sequences. Molecular and genetic approaches are combined to dissect this small region of the Drosophila chromosome.

We also want to correlate this detailed genetic map with other levels of chromosome organization : polytene chromosome banding, transcription maps and chromosomal scaffolding structures. We believe that correlations between different levels of organization may reveal structures which are essential for the control of the genetic activity.

THE POLYTENE CHROMOSOME

Polytene chromosomes found in the salivary glands of Drosophila larvae and in some other tissues arise by successive replications without chromosome separation. The giant interphase chromosomes exhibit a characteristic and constant banding pattern on which genes can be placed. In addition, the number of bands corresponds well to the number of essential genetic units estimated for the whole Drosophila genome (about 5000) and in most short regions analysed. We have firstly used the technique of in situ hybridization to localize DNA sequences from the chromosomal walk on the polytene chromosomes (Spierer et al. 1983). The 315 kilobase walk covers fifteen chromomeric units (band and adjacent interband). We have found that within our cloned interval, the genomic length of chromomeric units extends from 3 to 5 kilobases, for very faint bands, to 150 kilobases for a large one. In cytological terms, the large chromomeric units is not 30 to 50 times wider than the very small ones, but only about 10 times. Assuming that interbands have a similar sequence content in small and large chromomeric units, this indicates that the DNA is compacted in bands, as suggested by cytophotometry (Beermann 1972). Previous work, however was measuring DNA mass, and not genomic sequence length, leaving thus open the possibility that the high DNA content of bands was due to DNA over-replication. To end this old dispute, we have demonstrated directly that the level of DNA replication among bands and interbands of the polytene chromosome is monotonous in this interval (Spierer and Spierer 1984). Genomic blots were prepared from polytene DNA (salivary glands) and diploid DNA (young embryos). The relative abundance of over eighty restriction fragments spanning the walk was measured by densitometry of autoradiographs after hybridization to labelled recombinants. No significant difference in relative abundance of the restriction fragments could be found along the walk.

GENETIC ORGANIZATION

The region we study has been saturated for recessive lethal mutations by Hilliker and coworkers (1980). These authors have established a genetic map using a set of staggered deletions. We have mapped on the DNA the 16 breakpoints which disrupt the cloned interval (Spierer et al. 1983) as well as nine new breaks from deletions that we have generated recently (Gausz et al. 1986). These landmarks allowed to established a detailed genetic map at the molecular level. In addition to the 12 essential genetic units reported by Hilliker et al. (1980), three additional genes have been recently localized in the walk: snake, a gene required for the establishment of dorsal/ventral polarity in the embryo, hsc2, a gene homologous to the hsp70 heat-shock gene, and a suppressor of position-effect variegation. The genomic size of the genes in this interval is found to vary from a few kilobases to over 50 kilobases. One region contains five functionally independent genes in a 20 kilobase interval. The alignment of this genetic map with the polytene chromosome banding pattern map reveals that the regions with a high density of genes per DNA length are also dense in chromomeric units. The resolution of *in situ* hybridization did not suffice to establish one-to-one correlations between faint bands and genes. The large 150 kilobase long band, on the other hand, comprises one gene extending over at least 50 kilobases and a 45 kilobase region which can be deleted and still allows survival (it contains a gene which is semi-lethal).

GENETIC ACTIVITY

We have also established a map of the genetic activity, by detecting and characterizing 45 polyadenylated transcripts produced at various developmental stages and in some specialized tissues (Hall *et al*. 1983, Bossy *et al*. 1984, Hall and Spierer, unpublished, R. DeLotto, unpublished). The number of transcripts is hence three times larger than the number of detected genes. 63 kilobases of the walk (the region highly populated in genes and bands) encode 19 transcripts while the 150 kilobases covered by one large band and containing only two known genes encode only seven transcripts. Only six transcripts are found in all tissues and stages examined. These "housekeeping" genes are not clustered. The other transcripts are specific to one or a few tissues. The transcripts exhibiting a similar developmental profile are not clustered either.

CHROMOSOME SUPERSTRUCTURES

We have localized the attachment sites of the DNA to the nuclear matrix (or scaffold) in Drosophila tissue culture cells along the same interval (Mirkovitch, Spierer and Laemmli, 1986). We think that if the position of these sites is linked to the genetic organization, then they might play an essential role in the expression and replication of the genome. The nuclear scaffolds are prepared by extraction with LIS (Mirkovitch et al. 1984) and digested with restriction endonucleases. The scaffold-bound DNA is separated from the released DNA by sedimentation. The DNA sequence content of both fractions is then analyzed. We have detected four attachment sites along the 315 kilobases of genomic DNA. These attachment regions define looped domains of sizes extendig from 26 to 112 kilobases. The attachment sites do not interrupt the genetic units and they delimit domains containing one to up to five unrelated genes. The analysis of transcripts present in the same cultured cells indicates that expression of genes does not require immediate vicinity of a scaffold attachment regions since attachment sites can be separated from an active genetic unit by an unrelated gene inactive in these cells. We note however that the transcripts from this interval are expressed in low abundance and that association with an attachment site might be needed only for genes expressed abundantly.

PROSPECTS

This work has made this region the best known chromosome interval (of a relatively large size) in multicellular organisms. The accumulation of data from genetics, cytogenetics and molecular biology renders this region even more attractive for further studies, with interests ranging from technoloy developments to studies of evolution.

REFERENCES

Beermann, W. (1972). In Results and Problems in Cell Differenciation (Beermann, W. ed.), vol. 4, pp. 1-33, Springer Verlag (Berlin).

Bender, W., Spierer, P. and Hogness, D.S. (1983). J.Mol.Biol. 168, 17-33.

Bossy, B., Hall, L.M.C. and Spierer, P. (1984). EMBO J. 3, 2537-2541.

Gausz, J.,Hall, L.M.C., Spierer, A. and Spierer, P. (1986). Genetics 112, 65-78.

Hall, L.M.C., Masson, P.J. and Spierer, P. (1983). J.Mol.Biol. 169, 83-96.

Hilliker, A.J., Clark, S.H., Chovnick, A. and Gelbart, W.M. (1980). Genetics 95,95-110.

Mirkovitch, J., Mirault, M.E. and Laemmli, U.K. (1984). Cell 39, 223-232.

Mirkovitch, J., Spierer, P. and Laemmli, U.K. (1986). J.Mol. Biol. (in press).

Spierer, A. and Spierer, P. (1984). Nature 307, 176-178.

Spierer, P., Spierer, A., Bender, W. and Hogness, D.S. (1983). J.Mol.Biol. 168, 35-50.

Polytene chromosomes of *Drosophila*

C. Laird, M. Hammond and M. Lamb

Department of Zoology, University of Washington, Seattle, Washington 98195, USA

SUMMARY

In this review we discuss several aspects of the structure and ontogeny of Drosophila polytene chromosomes. Although most DNA sequences in euchromatic arms are uniformly replicated, there are exceptions that provide insights into chromosome structure. We discuss three exceptional sequences that are under-replicated and whose cytogenetic locations correspond to constrictions and ectopic fibers at their respective sites, which have been described previously as sites of "intercalary heterochromatin." These observations support the suggestion that "intercalary heterochromatin" is DNA with altered replication patterns rather than with the properties of sequence repetition or transcriptional inactivity that usually characterize centromeric heterochromatin.

The occurrence of under-replicated euchromatic sequences has led us to the concept of a relaxed DNA synthesis (S) phase, in which the normal constraints on reinitiation of DNA replication are not completely effective. Because of this relaxed S phase, and two other features of the differentiation pathway of endoreplication, we hypothesize that polyteny occurs by "default," or loss of functions normally used in the diploid cycle, rather than by gain of specialized functions. Thus, polytene chromosomes are viewed as being at the ground state of chromosome organization in endo-replicating cells. This hypothesis has developmental and evolutionary implications for the occurrence of polytene chromosomes in diverse tissues and organisms.

Chromosomes today, volume 9 Allen & Unwin 1987

EUCHROMATIN IS PREDOMINANTLY UNIFORM IN ITS REPLICATION

Renaturation kinetics of DNA from polytene nuclei of salivary glands and from diploid nuclei of embryos of Drosophila hydei indicate that most single copy sequences are uniform in their replication from diploidy to polyteny (Dickson, Boyd, and Laird, 1971). This general uniformity has been demonstrated also for D. melanogaster by quantitative Southern hybridization of cloned DNA sequences (Lifschytz, 1983; Spierer and Spierer, 1984). These experiments provide support for the general model of polytene chromosomes indicated in Figure 1, in which uniformly replicated single-copy sequences, which are predominantly or entirely in euchromatic arms, are contrasted with centromeric heterochromatin, ribosomal, and telomeric sequences, which are represented at reduced levels of polyteny (Laird et al., 1973).

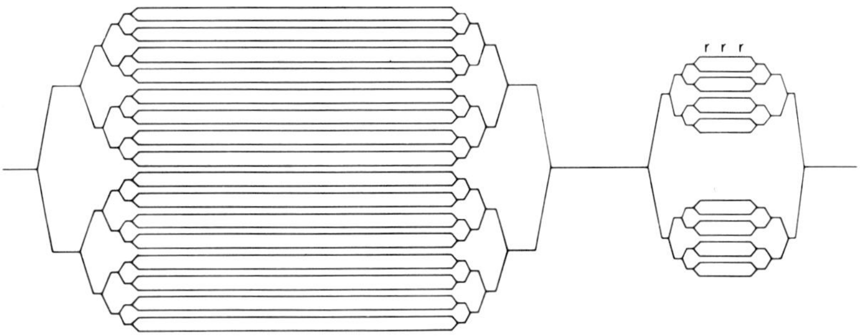

FIGURE 1. MODEL OF A POLYTENE CHROMOSOME (from Laird et al., 1973) Uniformly replicated sequences in euchromatic arms are contrasted with sequences reduced in replication levels, which include large ribosomal RNA genes (rrr), and centromeric heterochromatin on either side of the ribosomal genes. Still speculative is under-replication of telomeric sequences (chromosome ends).

EXCEPTIONS TO UNIFORM REPLICATION OF EUCHROMATIN: THREE UNDER-REPRESENTED DNA SEQUENCES ARE LOCALIZED IN CONSTRICTIONS, INDICATING A LIMITED CORRELATION BETWEEN REPLICATION AND CHROMOSOME MORPHOLOGY

The euchromatic arms of polytene chromosomes contain a few regions that are unusually wide or narrow. Based on densitometic analysis of high-voltage electron micrographs, we suggested that these regions of unusual width correspond to DNA sequences that are slightly advanced or retarded in their replication relative to most euchromatic DNA (Laird, 1980). No examples of over-replication in Drosophila polytene chromosomes, such as are observed in DNA puffs of Rhynchosciara (Breuer and Pavan, 1955; Glover et al, 1982), have yet been reported. However, three examples of under-represented DNA sequences have been analyzed. Spierer and Spierer (1984), mentioned that sequences in the Ubx region were under-represented in salivary gland DNA. Lifschytz (1983), presented data supporting a similar conclusion for histone gene sequences. We have analyzed these two examples, as well as a third case involving sequences from region 11A (Lamb and Laird, 1986). Using in situ hybridization and quantitative Southern analysis, we conclude that the under-represented sequences occur within the constriction, and, for Ubx and 11A, that adjacent and uniformly replicated sequences are localized next to but not within the constriction. Moreover, at least one of these regions, Ubx, contains a gradient of polyteny within a transcription unit, in which the 5′ region is four fold under-represented, while the 3′ region is uniformly represented, relative to DNA from diploid cells. Finally, another polytene tissue, fat body, does not exhibit either the under-representation of Ubx DNA (Lamb and Laird, 1986), or the constriction at 89E (Richards, 1980), that is present in salivary gland chromosomes. These correlations support the hypothesis that the morphologically observed constrictions are a consequence of the under-representation of sequences at these sites. Final proof of a causative role would require introduction of these sequences into ectopic chromosome sites, with the concomitant appearance of new constrictions and under-representation of introduced sequences at those sites.

DNA REPLICATION IS DELAYED BUT NOT COMPLETELY BLOCKED FOR SOME UNDER-REPRESENTED SEQUENCES

Hammond and Laird (1985b) showed by quantitative in situ hybridization that histone genes and ribosomal RNA genes are under-represented in polytene chromosomes of salivary glands, yet appear to participate fully in the final two to three rounds of endo-replication of these cells. Even a severely under-represented DNA sequence, the 1.705 gm/cm^3 satellite, shows almost a doubling in its abundance during these final endoreplication steps. Thus the inhibition of replication that leads to under-representation of these sequences must occur early in endoreplication, and be partially or completely relieved at a subsequent stage. We therefore refer to this as "delayed replication." This phenomenon has also been documented for another endoreplicating cell type of D. melanogaster. When nurse cells undergo their final endoreplication step from 756C to 1480C, histone seequences and the 1.705 gm/cm^3 satellite sequence double in their amount per nucleus (Hammond and Laird, 1985a). This result led to our view that delayed but active replication forks, as contrasted with blocked forks or discontinuous DNA, provide the basis for the transitions in polyteny (Figure 1). Multiple replication forks similar to those depicted in Figure 1 have been reported for another endoreplicating cell type, the follicle cells, during the stage at which amplification of chorion genes leads to a local transition in level of polyteny (Osheim and Miller, 1983).

THREE SITES OF UNDER-REPLICATION CORRESPOND TO SITES OF "INTERCALARY HETEROCHROMATIN"

Lamb and Laird (1986), point out that three sequences that are under-replicated in salivary glands are at or close to sites of "intercalary heterochromatin," which are regions in euchromatic arms with some cytogenetic properties in common with centromeric heterochromatin (see Zhimulev et al., 1982, for a summary of this topic). All three of these sequences (histone genes at region 39D, Pardue et al., 1977; Ultrabithorax (Ubx) at 89E, and a sequence at 11A, Lamb and Laird, 1986), are observed to be present in ectopic fibers at their respective regions. Thus by the criterion of ectopic pairing, these sequences correspond to "intercalary heterochromatin."

IS UNDER-REPLICATION OF EUCHROMATIC DNA SUFFICIENT TO GENERATE INTERCALARY HETEROCHROMATIN?

One of the strongest criteria of a site of "intercalary heterochromatin" is involvement in ectopic fibers that connect two non-homologous chromosomes or regions of a chromosome. It has been suggested that ectopic pairing fibers originate from branch migration of single strands at replication forks of under-replicated sequences (Ashburner, 1980). The single strands of DNA would be available either for pairing with other single strands of limited homology, or for an association with other single strands that is mediated by, for example, single-strand binding protein. In addition, the under-replication would be expected to result in a weak chromosomal region. (Weak sites are usually classified as "intercalary heterochromatin.") As summarized above, we confirm the predicted correlation of under-replication and ectopic fibers. However, the three sequences do not share two other major properties of centric heterochromatin: sequence repetition and general transcriptional inactivity. Ubx sequences are not repeated under the usual hybridization conditions, and normal amounts of histones are present in salivary glands (Cohen and Gotchel, 1971), presumably a result of transcription of the histone gene cluster.

Thus, the most striking feature in common among Ubx, histone gene, and 11A sequences in salivary glands is their pattern of delayed or under-replication. This result suggests that altered replication patterns of certain replication origins or termini are major factors leading to the observed properties of one important class of "intercalary heterochromatin" (Lamb and Laird, 1986). A corollary is that many euchromatic sequences may be classified as "intercalary heterochromatin" if they are within replication domains that are late replicating in salivary glands.

THE CHROMOSOMES OF ENDOREPLICATING NURSE CELLS ARE OLIGOTENE AND HAVE AN ORGANIZATION CLOSELY RELATED TO POLYTENY

Although nurse cells undergo as many cycles of endoreplication as do salivary gland cells in D. melanogaster, their chromosomes are not normally visible with the light microscope. We examined by in situ hybridization the replication patterns of specific sequences and the spatial arrangement of replicated chromatids in the nucleus (Hammond and Laird, 1985a). We found that the degree of association of homologous regions is locus specific (Fig. 2). In addition, nurse cell chromosomes exhibited a relaxed S phase as described above. These results indicate that nurse cell chromosomes may be considered to be polytene chromosomes that have dissociated to varying degrees along the chromosome arms (oligoteny; Ashburner, 1980). This view is consistent with a report that during the early stages of oogenesis, nurse cells have banded chromosomes (Brun and Chevassu, 1958). The similarity of oligotene and polytene chromosomes is underscored by the presence of mutations that lead to banded polytene chromosomes in nurse cells (see King, 1970; King et al., 1981).

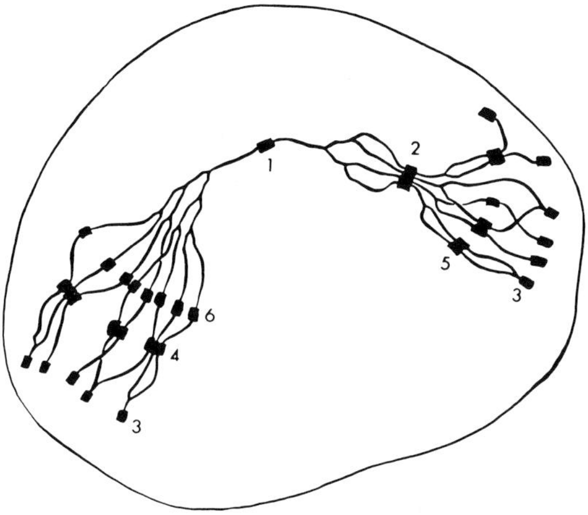

FIGURE 2. PROPOSED OLIGOTENE STRUCTURE OF NURSE CELL CHROMOSOMES (from Hammond and Laird, 1985a). Numbers refer to relative replication and pairing properties of a high-repeated centromeric sequence (1), a strongly pairing and moderately under-replicated histone gene cluster (2), a telomeric sequence (3), moderately paired and fully replicated sequences (4, 5), and weakly paired, fully replicated 5S RNA genes (6).

THE DEFAULT PATHWAY TO POLYTENY

The above considerations lead us to propose that polyteny is reached by a default pathway, in which several steps that normally occur in dividing cells are bypassed or interrupted. Endoreplicating cells bypass mitotic events of chromosome condensation, cytokinesis, and decondensation. They have a relaxed S phase, which we view as a loss of stringent control of DNA replication. In addition, there are mutations that lead to banded polytene, rather than oligotene, chromosomes in nurse cells (King, 1970; King et al., 1981). These mutations are fully recessive, which indicates that gene products are needed for the transition from oligoteny to polyteny. Thus we view polyteny as representing the ground state of chromosome structure in endoreplicating cells, rather than a specialized state.

If correct, the default pathway hypothesis has implications concerning the differentiation pathway leading to polyteny, and the evolution of polytene chromosomes in diverse organisms. The hypothesis also implies that polyteny can be induced in cells of other organisms, including human cells, by interference with steps that lead away from the ground state. This induction would provide chromosomes with considerably greater cytological resolution than is available presently. These implications are discussed in more detail elsewhere (Laird, 1986).

REFERENCES

Ashburner, M. 1980. Some aspects of the structure and function of the polytene chromosomes of the Diptera. In Insect Cytogenetics (Symposia of the Royal Entomological Society of London: number ten), R. L. Blackman, G. M. Hewitt, and M. Ashburner, eds. 65-84. London: Blackwell Scientific Publications.

Beermann, W. 1972. Chromomeres and genes. In Developmental studies on giant chromosomes, W. Beermann, ed., 1-33. Berlin, Heidelberg, New York: Springer-Verlag.

Breuer, M. E. and C. Pavan 1955. Behavior of polytene chromosomes of Rhynchosciara angelae at different stages of larval development. Chromosoma 7, 371-386.

Brun, J. and D. Chevassu 1958. L'evolution des structures chromatiniennes dans les cellules nourricieres de ovocytes chez Drosophila melanogaster Meigen. Chromosoma 9, 537-558.

Cohen, L. H. and B. V. Gotchel 1971. Histones of polytene and nonpolytene nuclei of Drosophila melanogaster. J. Biol. Chem. 246, 1841-1848.

Dickson, E., J. B. Boyd and C.D. Laird 1971. Sequence diversity of polytene chromosome DNA from Drosophila hydei. J. Mol. Biol. 61, 615-627.

Glover, D. M., A. Zaha, A. J. Stocker, R. V. Santelli, M. T. Pueyo, S. M. Detoledo and F. J. S. Lara 1982. Gene amplification in Rhynchosciara salivary gland chromosomes. Proc. Nat. Acad. Sci. USA 79, 2947-2951.

Hammond, M.P. and C.D. Laird 1985a. Chromosome structure and DNA replication in nurse and follicle cells of Drosophila melanogaster. Chromosoma 91, 267-278.

Hammond, M.P. and C.D. Laird 1985b. Control of DNA replication and spatial distribution of defined DNA sequences in salivary gland cells of Drosophila melanogaster. Chromosoma 91, 279-286.

King, R.C. 1970. Ovarian development in Drosophila melanogaster. New York, London: Academic Press.

King, R.C., S.F. Riley, J.D. Cassidy, P.E. White, and Y.K. Paik 1981. Giant Polytene Chromosomes from the Ovaries of a Drosophila mutant. Science 212,441-443.

Laird, C.D. 1980. Structural paradox of polytene chromosomes. Cell 22, 869-874.

Laird, C. D. 1986. Default pathway to polyteny. (in preparation).

Laird, C. D., W. Y. Chooi, E. H. Cohen, E. Dickson, N. Hutchison and S. H. Turner 1973. Organization and transcription of DNA in chromosomes and mitochondria of Drosophila. Cold Spring Harbor Symp. Quant. Biol. 38, 311-327.

Lamb, M. M. and C. D. Laird 1986. Three regions of "intercalary heterochromatin" in polytene chromsomes of Drosophila contain DNA sequences under-replicated and localized in constrictions (in preparation).

Lifschytz, E. 1983. Sequence replication and banding organization in the polytene chromosomes of Drosophila melanogaster. J. Mol. Biol. 164. 17-34.

Osheim, Y. N. and O. L. Miller, Jr. 1983. Novel amplification and transcriptional activity of chorion genes in Drosophila melanogaster follicle cells. Cell 33, 543-553.

Pardue, M. L., L. H. Kedes, E. S. Weinberg and M. L. Birnstiel 1977. Localization of sequences coding for histone messenger RNA in the chromosomes of Drosophila melanogaster. Chromosoma 63, 135-151.

Richards, G. 1980. The polytene chromosomes in the fat body nuclei of Drosophila melanogaster. Chromosoma 79, 241-250.

Spierer, A. and P. Spierer 1984. Similar level of polyteny in bands and interbands of Drosophila giant chromosomes. Nature 307, 176-178.

Zhimulev, I. F., V. F. Smemshin, V. A. Kulichkov and E. S. Belyaeva 1982. Intercalary heterochromatin in Drosophila. I. Localization and general characteristics. Chromosoma 87, 197-228.

Structure and function of Y chromosomal genes in *Drosophila*

W. Hennig, R. C. Brand, J. Hackstein, P. Huijser, C. Kirchoff, H. Kremer, D.-H. Lankenau and P. Vogt

Department of Genetics, Faculty of Sciences, University of Nijmegen, Toernooiveld, 6525 ED Nijmegen, The Netherlands

Our thinking about gene function is governed by the prejudice that genes have to code for proteins or, at least, for RNA which both are then functional in locations different from their genetic origin. We are also inclined to forget that nothing is known about the possible biological role of the majority of DNA sequences in eukaryotic genomes. It is clear that in most eukaryotes more than 90% of the DNA does not code for proteins or known species of RNA.

The studies on the molecular structure of the Y chromosomal male fertility factors of Drosophila hydei, carried out by our group over the past years, revealed that the biological role of these genes is apparently different from conventional gene functions. The molecular properties of these genes differ from those of other known genes and we propose that the biological function of the Y chromosomal genes is to accumulate nuclear proteins with the aid of the Y chromosomal transcripts. The arguments for this hypothesis are summarized in this lecture.

THE Y CHROMOSOME OF DROSOPHILA HYDEI

Contrary to the male sex chromosome of mammals, the Y chromosome of Drosophila is not involved in sex determination. In D. hydei, the species of our studies, it contains almost 10% of the genomic DNA. In somatic tissues it is not detectably functional but in the meiotic prophase of the male it becomes decondensed and active in transcription. A detailed description of the meiotic prophase of D. hydei has recently been completed (Kremer et al. 1986). Our

Chromosomes today, volume 9 Allen & Unwin 1987

genetic studies disclosed that the Y chromosome of D. hydei carries not more than 16 complementation groups. Five of them are related to the five lampbrush loops formed in the primary spermatocyte nucleus, each loop being associated with one complementation group (Hackstein et al. 1982). In spite of the high transcriptional activity of the Y chromosomal lampbrush loops we found no testis protein coded for by the Y chromosome (Hulsebos et al. 1983). To understand the genetic properties of the Y chromosomal genes we investigated the molecular properties of the DNA of several lampbrush loops. The results of earlier studies are summarized in a review (Hennig 1965) and the detailed results concerning the DNA structure in a series of recent publications (Vogt and Hennig 1986a, b; Vogt et al. 1986; Huijser and Hennig 1986).

Y CHROMOSOMAL DNA

Y chromosomal DNA sequences of D. hydei were isolated by three different methods, (1) a quantitative dot blot screen of recombinant plasmid DNA with labeled DNA from males or females (Vogt and Hennig 1983), (2) from a cDNA library from testis RNA (Brand et al. in preparation), and (3) by microcloning dissected lampbrush loops (Hennig et al. 1983).

Y chromosomal DNA sequences belong to two different classes of repetitive DNA sequences. One class is Y-specific, i.e. it is present only on the Y chromosome. The other class, Y-associate DNA, is with some copies present on the Y chromosome, other copies are located elsewhere in the genome (Vogt and Hennig 1983). Both classes of repeated DNA families have different properties and will be described separately.

Y-specific DNA sequences

Y-specific DNA sequences occur in repeat families of 200 - 2000 copies. They are arranged in clusters of tandem repeats of moderate length (200 - 400 bp) and are probably restricted in their location to single (Vogt and Hennig 1986a) or functionally related (Huijser and Hennig 1986) Y chromosomal lampbrush loops. DNA sequence analysis gives no evidence for protein coding properties in such sequences. The sequence character, however, displays a number of other remarkable features. First, the sequences have a high capacity to form secondary structures due to internal direct and inverted sequence repeats. In some sequence families of Y-specific DNA, sequences with homology to ARS sequences of yeast and to the DNA sequences supposed to be involved in the amplification mechanism for the chorion genes in Drosophila follicle cells are present. Also a high degree of homology to enhancers of transcription was found (Vogt and Hennig 1986a, b). We assume that these features are not accidental but they are connected to the function of this DNA sequence family.

We studied the evolutionary conservation of two Y-specific DNA sequence families. Both are not highly conserved in evolution, since they are either restricted to few closely related species or even absent in species other than D. hydei. This observation agrees with the high degree of species specificity of the lampbrush loop morphology. We assume that the presence of DNA sequences homologous to ARS elements is important for the creation of new Y-specific DNA sequence families by amplification events (for discussion see Vogt et al. 1986 and Huijser and Hennig 1986).

Y-associated sequences

The copy numbers of members of Y-associated DNA sequence families are smaller than those of members of Y-specific DNA sequence families. They may include only a few copies and seem usually not to exceed 50 copies. The length of the DNA sequences in this class of repeated sequences is highly variable and may range up to several kb. Our general picture of the properties of these DNA sequences is still rather incomplete.

The designation of this class of repetitive DNA sequences was chosen because only few of the copies are Y chromosomal. In DNA blots it is often difficult to establish the Y chromosomal location of such sequences. But in situ hybridization experiments clearly demonstrate the Y chromosomal location.

Also for Y-associated DNA sequences the nucleotide sequences give no evidence for a protein coding function (Vogt and Hennig 1986b). The sequence properties reveal a potential to form secondary structures in transcripts, comparable to the features of the Y-specific DNA sequences.

The most intriguing character of the Y-associated DNA class is our observation that these sequences display the properties of transposable DNA sequences. A comparison of the autosomal and X chromosomal location of Y-associated DNA sequences in closely related species of D. hydei (Hennig et al. in preparation), or even in different strains of D. melanogaster (Lankenau et al., unpublished), shows that the non-Y chromosomal copies of this DNA class are highly variable in their genomic location.

The pattern of evolutionary conservation of Y-associated DNA sequences differs from that of Y-specific DNA sequences. We observed that Y-associated DNA sequences are evolutionarily much more conserved and occur also in species phylogenetically more distant to D. hydei (Vogt et al. 1986). It is of particular interest that in these distant species no Y chromosomal location of DNA sequences from the Y-associated class was found. This suggests that Y-associated DNA sequences are only secondarily inserted into their Y chromosomal positions due to their transposable character (for discussion see also Hennig 1986).

Our considerations must now be directed to the question which local relationship between Y-specific and Y-associated DNA sequences exists. Nucleotide sequence analysis revealed that both types of DNA occur closely linked (Vogt and Hennig 1986b). The Y-associated DNA sequences are interspersed between clusters of Y-specific DNA sequences. Each cluster of Y-specific DNA sequences is probably characterized by specific sequence features caused by sequence divergence and subsequent amplification (see Vogt and Hennig 1986a, b).

Y CHROMOSOMAL TRANSCRIPTS

Already in 1968 it was demonstrated with the aid of hybridization experiments that Y chromosomal repetitive DNA sequences are in testes represented in transcripts. The difficulties to assess the actual genomic origin of transcripts of repeated DNA sequences were pointed out (Hennig 1968). With the aid of more recent techniques it is possible to obtain more detailed informations on the transcription of repeated DNA sequences.

Blots of testis RNA after electrophoresis and hybridization with labeled cloned Y chromosomal DNA sequences showed that both classes of DNA, Y-specific and Y-associated sequences, are represented in testis-specific transcripts. Important information on the mode of transcription of Y chromosomal lampbrush loops was obtained from Miller spreading experiments. They revealed that within the loop giant primary transcripts are formed which are display a high degree of secondary structure. The size of the transcripts ranges between 260 and more than 1000 kb, dependent on the particular loop. The DNA of the loops is, therefore, at least for a large part transcribed. The morphology of the transcripts is loop-specific which finds an equivalent in the loop-specificity of the DNA sequences and their capacity to form characteristic secondary structures (Grond et al. 1983;de Loos et al. 1984).

The demonstration by electron microscopy that the lampbrush loops are transcribed into giant primary transcripts implies that also Y-associated DNA sequences must occur in these molecules. By biochemical techniques it would be difficult to prove their transcription within Y chromosomal sites. An independent proof of the occurrence of transcripts of Y-associated DNA sequences within loops has been achieved by transcript-in situ hybridization. In such experiments the presence of transcripts within the loops is directly evident.

The hybrid patterns after size separation of testis RNA on acrylamide gels are substantially different for RNA sequences derived from Y-specific or Y-associated DNA sequences. While all Y-specific transcripts studied so far display a pattern of heterogeneous sizes (Vogt et al. 1982, and unpublished data), RNA species homologous to the Y-associated DNA class occur in discrete size classes (Brand et al. 1985, and in preparation). The biological role of such RNA species remains unclear. We have not suceeded to obtain proteins from such RNA species isolated by hybrid-select techniques in an in vitro translation system (Brand, Huijser, W. Hennig and B. Hennig, unpublished data).

It is of particular interest that part of the Y-associated DNA sequences in non-Y chromosomal locations must be transcribed. In testis RNA of X/O males transcripts homologous to such Y-associated DNA sequences were discovered (Brand et al. in preparation). Further studies of such sequences might therefore be suited to provide evidence on the biological relevance of the Y-associated DNA class. We suggested that the homology between autosomal and Y chromosomal sites might be required for regulatory interactions (Vogt and Hennig 1986b; Kremer et al. 1986).

FUNCTION OF Y CHROMOSOMAL GENES

We have seen that the character of Y chromosomal DNA sequences makes it unlikely that much of this DNA codes for proteins. Although we cannot exclude that minor fractions of the loop DNA are involved in protein coding but have so far escaped our attention, such sequences cannot code for amounts of proteins as would be expected to be synthesized according to the transcriptional activity in Y chromosomal lampbrush loops. Moreover, the molecular structure of the Y chromosomal genes finds no comparable equivalent in other protein coding genes studied so far. This has induced us to search for alternative gene functions assigned to the Y chromosomal genes.

An example of a Y-specific DNA-sequence with protein-binding capacity

As mentioned before, the transcripts of Y-specific DNA sequences have the capability to form highly complex secondary structures. The nucleotide sequence analysis also revealed that the Y chromosomal sequences have properties as typically found in nucleic acid sequences interacting with proteins (see Vogt and Hennig 1986a, b). This is particularly well examplified by a Y-specific DNA sequence (designated as "rally" sequence: ribosomal and lampbrush loop Y chromosomal sequence) recovered by microcloning (Huijser and Hennig 1986). After microdissection of the lampbrush loop "pseudonucleolus" from the distal part of the long arm of the Y chromosome we identified a tandemly repeated, 193 bp long DNA sequence which is homologous to a part of the 26S rRNA molecule 600 bp downstream the IVS insertion site. This sequence is highly conserved between different organisms. In bacteria it contains the binding site for the ribosomal protein L6. In situ hybridization showed that this Y chromosomal sequence resides in the two lampbrush loop pairs "threads" and "pseudonucleolus" distal in the short arm and is present in transcripts of these loops. In the context of this lecture it is impossible from time reasons to discuss this rather interesting sequence in detail. However, in discussing potential functions of the lampbrush loops this sequence is particularly well suited to emphasize the possible role of Y chromosomal transcripts as protein binding sequences.

Our analysis of the rally sequence emphasizes that the molecular structure of Y chromosomal genes is even more complex than indicated by the data mentioned until now. The rally sequence occurs not only in two adjacent lampbrush loop pairs but it is also restricted to a specific region of the "threads" close to the nucleolus. Our Miller spreading experiments have shown that in this part of the "threads" a complex secondary structure is formed by the transcripts while the more distal region of the loop display a different transcript structure (see deLoos et al. 1984). This bipartite moleculare structure could now also be confirmed by molecular data. Further data (Huijser and Hennig, unpublished) revealed that between the lampbrush loops "threads" and "pseudonucleolus" a transcribed poly(CA) repeat is located. It is correlated with the "cones", structures associated with the "pseudonucleolus". Such CA-rich nucleic sequences seem generally to be important in the context of protein binding. Moreover, the sequence homology between the "threads" and "pseudonucleolus" suggests an evolutionary and possibly also functional relationship between both loop pairs. A complex interaction between these lampbrush loop pairs was already earlier concluded from genetic data (Hackstein et al. 1982; Hulsebos et al. 1984; Hackstein et al., in preparation).

Molecular composition of the lampbrush loops

The idea of binding of specific proteins to the Y chromosomal lampbrush loops is supported if the molecular composition of these loops is considered. The DNA concentration in the Y chromosomal lampbrush loops is extremely low. Only recently it has been possible to demonstrate the location of the DNA axis in some loops with a fluorescent dye binding to DNA (Kremer et al. 1986). Although the loops are rich in RNA the major part of the loop matrix consists of protein (Grond et al. 1984). In particular those loop regions highly refractive in phase contrast, consist preferentially or exclusively of protein. Immunological studies revealed the presence of different types of proteins. Proteins related to proteins in RNP were demonstrated to occur distributed over all loops (Glätzer 1985). But also proteins with a high degree of loop-specificity were discovered (Hulsebos et al. 1984; Hennig et al. unpublished data; see also Hennig 1985). Such proteins may include histones or histone-related proteins (Kremer et al.1986).

The occurrence of loop-specific proteins raises the question in which way the loop-specific assembly of these loop components is achieved. In earlier parts of this lecture arguments in favour of protein binding capacities of lampbrush loop transcripts were derived from the nucleotide sequence properties. These observations now allow to complete our picture of the loop functions. The - loop-specific - secondary structure of the transcripts may permit to select for specific proteins which are subsequently stored along the loops. Although some arguments favour the possibility that the protein storage occurs in association with RNA, our earlier electron micorscopy studies did not provide much evidence for the presence of nucleic acids in the loop regions refractive in phase contrast microscopy (Grond et al. 1984). We can, however, not exclude that part of the proteins in association with RNA leave the spermatocyte nucleus.

Potential biological functions of lampbrush loop proteins

The protein assembly in the lampbrush loops may be related to different biological functions. Evidence for one function has been obtained from cytological studies (Kremer et al. 1986). The detailed investigation of the chromatin constitution during spermatogenesis revealed complex chromatin rearrangements, in particular during the early postmeiotic phase. In X/0 males, the chromatin rearrangements are abnormal (Kremer, unpublished observations). This may be considered as an indication for the involvement of loop proteins in chromatin rearrangements during spermatogenesis.

A more direct argument is derived from our observation that the monoclonal antiserum Bv96/2 (kindly provided by Dr. M. Frasch, Tübingen) reacts with the lampbrush loop pair "nooses" in the short arm of the Y chromosome. It also reacts with antigens in the sperm heads of almost mature spermatids (Hennig, v.d. Heide, Hendrikx, unpublished observations). We, therefore, suppose that the lampbrush loop "nooses" binds chromosomal proteins which are in later stages of spermatogenesis constitutents of the (inactive) chromatin.

Possible alternative functions of the loop transcripts and proteins were recently discussed by Vogt and Hennig (1986b) and Kremer et al. (1986). Such functions are suggested by our observations that (1) loop-specific transcripts are present in the cytoplasm of spermatids (the occurrence of Y chromosomal transcripts in postmeiotic stages had already been demonstrated)and that (2) DNA sequences from the Y-associated class located in non-Y chromosomal sites are transcribed in testes.

Contrary to many differentiation processes in eukaryotes, the postmeiotic morphogenetic development of male germ cells cannot be regulated by differential gene activity since RNA synthesis ceases during the first meiotic prophase. Morphogenesis must hence be subject to different modes of regulation. It is likely that self-assembly processes play an important role. But also such processes must be initiated and controlled. We assume that the lampbrush loop transcripts and their associated proteins are involved in the control of these events. A regulatory role of the Y chromosomal genes is suggested by the relatively unspecific effects in spermiogenesis caused by Y chromosomal mutations. Other supporting evidence for a regulatory role of the Y chromosome was provided by our studies of testis proteins. We found that three major protein fractions (sph35, tubulins and sph155) from testes in their expression are dependent on the presence of the Y chromosome (Hulsebos et al. 1984) although their genomic origin is non-Y chromosomal. More recent experiments disclosed that at least the expression of the tubulins is controlled at a post-transcriptional level (Brand et al. in preparation).

The complexity of these various observations implies that the function of the Y chromosomal genes can be expected to concern different levels of the development of a male germ cell. Our studies provide the opportunity to experimentally approach some of these functions.

CONCLUSIONS

The observations reported in this lecture have induced us to propose a new function for the Y chromosomal fertility genes of Drosophila. We assume that the primary function of these genes is to collect distinct proteins during the meiotic prophase which are subsequently required in morphogenesis of the male germ cell. Also Y chromosomal transcripts may be involved in regulatory processes which occur after meiosis. It has been pointed out earlier that such a gene function as postulated for the Y chromosomal fertility genes is not necessarily an exclusive character of spermatogenesis in Drosophila, but that genes of comparable functions may be expressed in other differentiation pathways and organisms. Protein binding by active genes might, for example, be used to regulate the expression of other sets of genes or for other regulatory processes. Examples for other genes with similar functions may be some of the heat shock genes, which in their molecular structure are in many features similar to the Y chromosomal fertility genes.

Finally, it should be mentioned that the molecular structure of Y chromosomal DNA are of general interest for understanding the structure and function of heterochromatin. We think that the molecular structure of the Y chromosome of D. hydei closely resembles general features of heterochromatin (see Hennig 1986).

REFERENCES

Glätzer, K.H. (1984). Preservation of nuclear RNP antigens in male germ cell development of Drosophila hydei. Mol. Gen. Genet. 196: 236-243.

Grond, C.J., Rutter, R.G.J., Hennig, W. (1984). Ultrastructure of the Y chromosomal lampbrush loops in primary spermatocytes of Drosophila hydei. Chromosoma 89, 85-95.

Hackstein, J.H.P., Leoncini, O., Beck, H., Peelen, G., Hennig, W. (1982). Genetics fine structure of the Y chromosome of Drosophila hydei. Genetics 101, 257-277.

Hennig, W. (1985). Y chromosome function and spermatogenesis in

Drosophila hydei. Adv. Genet. 23, 179-234.
Hennig, W. (1986). Heterochromatin and germ line-restricted DNA. Results and Problems in Cell Differentiation, vol 13 (W. Hennig, ed.). Springer-Verlag, Heidelberg.
Hennig, W., Huijser, P., Jäckle, H., Edström, J.-E. (1983). Molecular cloning of microdissected lampbrush loop DNA sequences of Drosophila hydei. The EMBO J. 2, 1741-1746.
Huijser, P. Hennig, W. (1986). Ribosomal DNA-related sequences in a Y chromosomal lampbrush loop of Drosophila hydei. submitted for publication.
Hulsebos, T.J.M., Hackstein, J.H.P., Hennig, W. (1983). Involvement of Y chromosoma loci in the synthesis of Drosophila hydei sperm proteins. Dev. Biol. 100, 238-243.
Hulsebos, T.J.M., Hackstein, J.H.P., Hennig, W. (1984). Lampbrush loop-specific protein of Drosophila hydei. Proc. Nat. Acad. Sci. US 81, 3404-3408.
Kremer, H., Hennig, W., Dijkhof, R. (1986). Chromatin constitution in the male germ line of Drosophila hydei. Chromosoma, in press.
de Loos, F., Dijkhof, R., Grond, C.J., Hennig, W. (1984). Lampbrush loop-specific transcripts in Drosophila hydei. The EMBO J. 3, 2845-2849.
Vogt, P., Siegmund, I., Hennig, W. (1982). Identification of cloned Y chromosomal DNA sequences from a lampbrush loop of Drosophila hydei. Proc. Nat. Acad. Sci. US 79, 5132-5136.
Vogt, P., Hennig, W. (1983). Y chromosomal DNA of Drosophila hydei. J. Mol. Biol. 167, 37-56.
Vogt, P., Hennig, W. (1986a). Molecular structure of the lampbrush loop nooses of the Y chromosome of Drosophila hydei. I. A Y chromosome-specific repetitive DNA sequence family is dispersed in the loop DNA. Submitted for publication.
Vogt, P. Hennig, W. (1986b). Molecular structure of the lampbrush loop nooses of the Y chromosome of Drosophila hydei. II. Non-Y specific DNA sequences are a major constituent of the loop. Submitted for publication.
Vogt, P., Hennig, W., ten Hacken, D., Verbost, P. (1986). Evolution of Y chromosomal lampbrush loop DNA sequences of Drosophila. Chromosoma, submitted for publication.

Acknowledgements.

We are grateful to Dorette ten Hacken, Rosilde Dijkhof, Mariette Dircks and Wieli Janssen for expert technical assistance, and to Marijke Smits for typing the manuscript.

CHROMOSOMES AND EVOLUTION

A model for heterochromatin dispersion and the evolution of C–band patterns

D. Schweizer and J. Loidl

Institut für Botanik der Universität Wien, Rennweg 14, 1030 Wien, Austria

SUMMARY. Taking many observations from many species an attempt was made to find general regularities and rules governing the karyotype position of C-bands in order to provide clues to the mechanisms of cytological and molecular evolution of patterns. A model is presented to explain the identified rules.

Within a heterochromatin-rich complement, the shorter chromosomes (or chromosome arms) tend to have large telomeric C-bands whereas the longer chromosomes tend to have intercalary C-bands. The intercalary C-band distribution patterns are in many species a true reflection of the spatial disposition of the telomeres in the mitotic interphase nucleus (Rabl polarisation). Heterochromatin formed at telomeres may from there be transferred to equilocal interstital sites in non-homologous chromosome arms. Transfer requires proximity. Proximity is determined by chromosome disposition in the three-dimensional space of the interphase nucleus. Our model assumes, on the basis of the identified rules, that the molecular processes of heterochromatin amplification and transposition act predominantly at *mitotic* interphase.

Our model further assumes that the mechanisms responsible for conservation of differences between C-bands and for the homogenisation of heterochromatic DNA sequences respectively act predominantly during first *meiotic* prophase. This notion is based on a comparison of mitotic and meiotic spatial nuclear organisations (Rabl polarisation vs. bouquet polarisation) in Robertsonian systems. Recombination requires proximity. Proximity is determined by spatial chromosome disposition. We suggest that the nuclear organisation at zygotene and pachytene is a major determinant.

In sum it appears that the incipient evolution of C-band patterns is constrained by nucleotypic properties and the chromosome arrangement at mitotic interphase. The evolution of biochemical differences of equilocal C-bands may be constrained by the nuclear organisation at early stages of first meiotic prophase.

INTRODUCTION TO THE MODEL

Regularities of C-band distribution, which are described in the following suggest a mechanism for the evolution of C-band pattern, which operates above the level of the individual chromosome arm and which seems to be a common feature in a wide range of organisms.

We propose a model for such a mechanism which provides a rationale for the phenomena of equilocal C-band distribution, the coevolution of C-band patterns in similar chromosome arms and the concerted evolution of heterochromatic sequences in similar chromosomal positions.

The model is based on three kinds of *molecular processes*:

1) mechanism(s) for sequence amplification
2) mechanism(s) for sequence transfer
3) mechanism(s) for sequence homogenisation.

Interchromosomal sequence transfer and recombination requires proximity. Proximity is determined by the chromosomal disposition in the mitotic and meiotic nucleus. The model is therefore based on two kinds of *nuclear constraints*:

1) in the mitotic nucleus, exerted by the relic anaphase disposition of chromosome arms, i.e. the Rabl polarisation, and
2) in the early meiotic (zygotene/pachytene) nucleus, exerted by the disposition of membrane attached bivalents, i.e. the bouquet polarisation.

REGULARITIES IN C-BAND DISTRIBUTION PATTERNS

Studies on C-band distribution have been carried out mainly on two groups of organisms, (1) in acridoid grasshoppers (e.g. John & King 1977, 1980, 1983; King & John 1980; John et al. 1985) and (2) liliaceous plants (e.g. Bentzer & Landström 1975; Vosa 1979; Blakey & Vosa 1981, 1982; Greilhuber et al. 1981; Ambros 1983; Loidl 1983). Both groups contain species with relatively large chromosomes and discrepant amounts of heterochromatin; in both groups too, heterochromatin polymorphisms are frequent within populations.

From these investigations, as well as from banding studies in several other organisms, the following regularities and restrictions governing the distribution of C-bands became apparent:

1) **Basic heterochromatin endowment**: When present, C-bands are preferentially located at the telomeres, the centromeres and at the nucleolus organizing regions.

2) **Banding style** (Greilhuber & Speta 1976; Schweizer & Ehrendorfer 1976): All chromosomes within a complement do usually show similar banding characteristics (i.e. distribution, amount and base composition), which deviate only in chromosomes with special functions (see below).

3) **Band pattern and chromosome size**: Pronounced telomeric bands tend to be preferentially associated with short chromosomes or chromosome arms, whereas the longer chromosomes or chromosome arms of a complement tend to develop intercalary bands (Fig. 1).

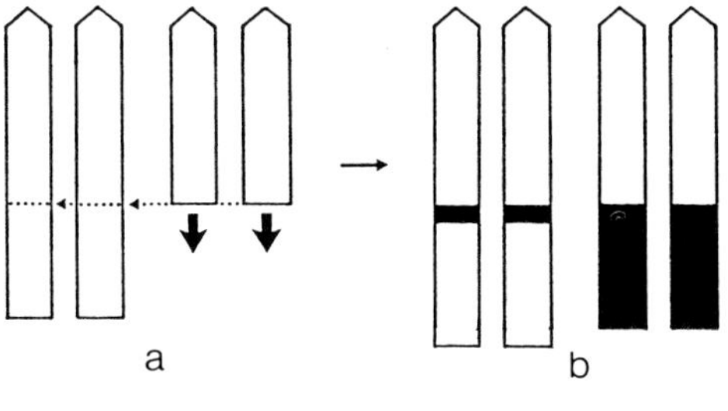

Fig. 1. Behaviour of repetitive terminal sequences: transposition to equilocal intercalary regions in longer chromosome arms and/or terminal accumulation in shorter chromosome arms (**a**). The gain of C-band material in shorter arms is usually greater than in longer arms (**b**).

4) **Equilocal heterochromatin distribution**: Constitutive heterochromatin (C-band material) in non-homologous members of the same diploid set tends to be located at similar sites with respect to the centromere-band distance ('Äquilokale Heterochromatie' - Heitz 1933).

5) **Regularities and restrictions governing the positions of intercalary bands**: For each intercalary band, there exists a chromosome arm within the population, whose telomere has the same distance from the centromere as the proximal border of that band. The telomeres of the shorter chromosome arms thus define the positions of the intercalary bands in the longer chromosome arms and the shortest chromosome arm of a population roughly determines a pericentromeric region with little or no banding (Loidl 1983).

These relationships of chromosome length with the location of intercalary C-bands suggest a role for telomeres in C-band initiation (Greilhuber & Loidl 1983; for discussion see Schweizer & Ehrendorfer 1983, John et al. 1985).

Chromosomes with the centromere in a median or submedian position tend to exhibit symmetrical banding patterns in their two arms (Greilhuber 1984).

DISCUSSION OF THE MODEL

Remarks on the terms and descriptions used. The terms used in this paper are meant to be non-commital descriptions of certain cytological phenomena and their possible molecular mechanisms. Telomeric heterochromatin denotes C-band material located at the very terminal site of a chromosome arm. At the cytological level the resolution of such a band is about 10^6 bp - 10^7 bp and hence, no further distinction is made between truly telomeric DNA sequences and subterminal sequences. 'Heterochromatic DNA' (Brutlag 1980) is equivalent to tandemly arranged highly repeated DNA sequences or satellite DNAs (see Singer 1982).

The term heterochromatin amplification is used to denote an increase in amount of constitutive heterochromatin. The term 'sequence amplification' refers to an increase in copy number of a DNA sequence. The term 'amplification' does not imply this increase was 'sudden' (e.g. by saltatory replication, disproportiate replication), or 'slow' (e.g. by unequal exchange), or whether or not the amplification process was coupled with

transposition (for discussion see Smith 1976, Kurnit 1979, Bird 1980, Flavell 1982).
Terms such as transposition, spread, transfer and dispersion of heterochromatin are used to denote the transfer of sequence information contained within heterochromatic DNA at a particular site (and hence the respective cytologically defined properties of this C-band material) to a new site in the karyotype. This transfer is probably replicative rather than conservative, because, as will be discussed in our model, the donating C-band sites usually remain and may even further amplify, after the C-band 'information' is conveyed to a new site.

The role of mitotic chromosome disposition in the evolution of C-band patterns: The Rabl-polarisation of the mitotic nucleus

We believe that the phenomena of C-band equilocality and the influence of armlengths on the location of bands are both a reflection of specific neighbourhood relationships of chromosomes within the interphase nucleus. It is widely acknowledged that the spatial disposition of the chromosomes at mitotic interphase has relic features of late ana- and telophase. The so-called 'Rabl-orientation' or 'Rabl-polarisation' of chromosomes (Rabl 1885; see Wilson 1925) describes the orientation of the chromosomes, with the centromeres directed to the spindle pole of the preceding mitosis. It is maintained through interphase to the following mitotic prophase. This persistence of the telophasic chromosome arrangement at interphase is probably secured by the attachment of the telomeres to the nuclear envelope (see Comings 1980).

Equilocal C-band distribution. A model which attempted to explain the concept of equilocality (Heitz 1933; Loidl 1983) assumes that telomeres (including 'centric' regions in the acro-telocentric chromosomes) are C-band initiation sites (Greilhuber & Loidl 1983; Schweizer & Ehrendorfer 1983). These regions are presumptive sites, where heterochromatin amplification tends to be initiated and from which heterochromatin sequences are dispersed. Transposition (transfer) may be coupled with the amplification process. Transpositional spread requires proximity and/or physical contact

(Arnheim 1983; Flavell 1982; Barnes et al. 1985). All the potential contact zones between a given telomere and those intercalary chromosome regions which lie at a similar distance from the centromere (i.e. 'equilocal' sites) are potential sites of heterochromatin transfer (from telomere to interstitial site) during evolutionary time periods (Fig.1). The model assumes that this kind of transfer is effective in evolution in mitotically dividing cells. As will be discussed below, the mechanisms responsible for the maintenance of sequence homogeneity may well be restricted to the meiotic process.

The model thus predicts that a karyotype consisting of chromosomes with approximately similar arm length will tend to have telomeric C-bands rather than intercalary bands. A karyotype consisting of members with different sizes and arm ratios, on the other hand, will also tend to possess intercalary bands in addition to telomeric bands.

Coevolution of C-band patterns in similar chromosome arms. A matching of the C-bands and their composition between similar chromosome arms was observed in *Secale africanum* and in *S. cereale* (Bennett 1982). A striking example for the coevolution of C-bands are the C-band patterns of gradually increasing complexity in the phylogenetic sequence *Scilla vindobonensis* (15 major bands, 10.6 % heterochromatin), *S. voethorum* (18 major bands; 15.4 % heterochromatin) and *S. resslii* (27 major bands; 22.1 % heterochromatin) (Greilhuber et al. 1981). Here too, progressive matching of C-band patterns between chromosome arms of similar size can be observed (Greilhuber & Loidl 1983). This phenomenon is probably again an expression of mitotic nuclear constraints. For purely mechanical reasons the telomeres of similar chromosome arms may have the highest chance of meeting and forming temporary associations. Concerted C-band evolution in similar non-homologous arms, such as is found in *Scilla* and in *Secale*, could therefore simply be a reflection of their statistically more frequent association (Fussell 1984) rather than reflecting a genetically fixed specific chromosomal order in the interphase nucleus (Ashley & Pocock 1981;

Bennett 1982; Lavania & Sharma 1985). A similar interpretation has recently been expressed by Barnes et al. (1985).

C-band heterogeneity. The causes of heterogeneity in C-band composition may be manyfold and yet they can be consistent with our model.

The present day distribution of C-bands probably does not reflect the result of a single transpositional sequence amplification event in the past. Different kinds of C-bands may have different origins in terms of telomeric C-band initiation sites and may have evolved at different times (see Young et al. 1983). Some sequences may have arisen by introgression from genomes of other races or subspecies.

Whether different repetitive sequences do co-evolve and become homogenized or whether they retain their heterogeneity may also be determined by factors which operate at meiosis.

The role of meiosis in the evolution of C-band patterns

Any general model for the evolution of C-band patterns in sexually reproducing organisms has to include features associated with the meiotic process. Models based solely on somatic chromosome mechanics will not suffice.

As shown recently by Charlesworth et al. (1986), increased rates of exchange will speed up the process of stochastic loss of repeated sequences. Therefore, a component of the model of Charlesworth and his coworkers (lit. cit.) is the coincidence of sites harbouring highly repeated DNA sequences and regions of restricted crossing over. This feature of their model is supported by cytological and genetic data. However, if regions in homologous chromosomes were completely devoid of recombinational processes, then satellite DNAs would be expected to accumulate and diverge over evolutionary time as is exemplified by the heterochromatic DNA of the XY sex chromosomes and the heteromorphic arm regions of chromosome 1 in newts (Sims et al. 1984). Clearly this is not the case for the majority of heterochromatic DNAs. Therefore, recombinational mechanisms including non-reciprocal exchanges such as gene conversion, operating between C-band DNA of homologous and also between heterochromatin of non-homologous chromosomes have to be involved for the concerted evolution or homogenisation of satellite DNA (see also the discussions by Arnheim et al. 1980, Dover & Flavell 1984).

Meiotic bouquet polarisation. We have already highlighted the first important feature of our model, namely the chromosomal Rabl-orientation at mitotic interphase and the transpositional amplification of sequences spreading from telomeres to equilocal positions in terms of centromere-telomere distance. The second important feature of our model is the operation of processes responsible for the maintenance of satellite DNA sequence homogeneity at early meiotic prophase. Here too, an essential component is the physical association of heterochromatic regions. Physical contact at meiosis is facilitated by the proximity of C-bands of heterologous bivalents, such as the telomeric clustering at the so-called bouquet stage during the early stages of synapsis.

Robertsonian systems. The importance of bouquet polarisation at meiosis is especially clear in the case of Robertsonian systems involving the fusion of two acrocentric (telocentric) chromosomes to form a single biarmed chromosome or else the dissociation of a biarmed chromosome into two telocentrics (see Holmquist & Dancis 1980).

Fig. 2 shows that the disposition of acrocentric chromosomes is not in any way altered at mitotic ana-/telophase by a Robertsonian event (fusion or dissociation). By contrast, the disposition of bivalents at meiotic prophase is significantly altered. Centromeric regions, which in telocentrics are clustered at the nuclear membrane during zygotene are expected to be spaced out when involved in a fusion event. This difference provides a rationale for the frequently observed differentiation of centromeric heterochro-

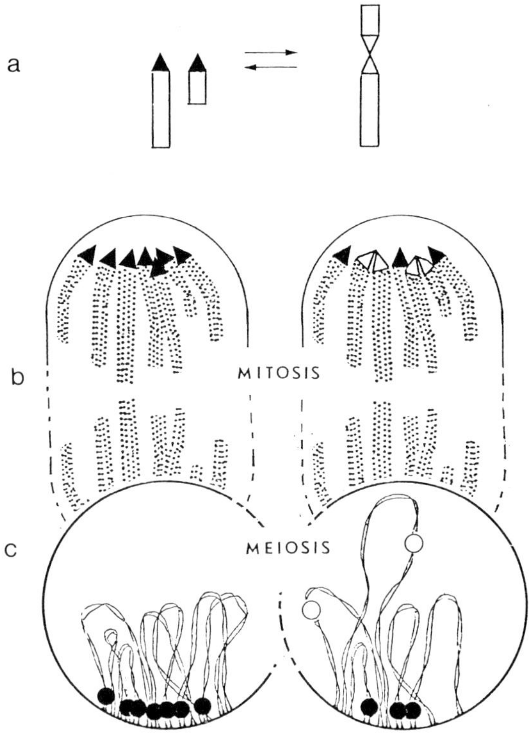

Fig.2. A Robertsonian event (**a**) does not alter the disposition of the centromeres in mitosis (**b**) whereas in meiotic prophase (**c**) the centromeres of (sub-)metacentric chromosomes are separated from the others during bouquet formation. This would allow the centric heterochromatin of the biarmed chromosomes to evolve independently if mechanisms of sequence homogenisation would require proximity and act during early meiosis.

matin in biarmed and telocentric chromosomes in a given Robertsonian system.

The proximity of heterochromatic DNA in the zygotene and pachytene nucleus, i.e. at stages when the enzymatic machinery for both reciprocal and non-reciprocal recombination is available, may provide a cytological basis for those molecular processes which are responsible for the maintenance of satellite DNA sequence homogeneity between groups of C-bands on both homologous and nonhomologous chromosomes.

Sus scrofa has a karyotype structure resembling the situation encountered in Robertsonian systems (Schwarzacher et al. 1984, Hamilton & Schweizer, unpublished). Other domestic animals, in particular some species belonging to the *Bovidae*, such as the species pair sheep and goat, are characterized by Robertsonian systems. In sheep the three largest chromosomes, which are Robertsonian products, contain sheep DNA satellite II but little or no satellite I sequences whilst the centromeric C-bands of the remaining unfused acrocentric autosomes contain sequences homologous to both satellites. In the goat, which has a presumed 'ancestral' karyotype consisting solely of telocentric chromosomes, all centromeric regions harbour satellite DNAs I and II (Kurnit et al. 1978).

An additional example of a change of C-band properties in conjunction with a Robertsonian mutation has been described for the orchid species pair *Cephalanthera damasonium* and *C. longifolia* (Schwarzacher & Schweizer 1982).

Apparent exceptions to the model

There are a number of species wherein the C-band patterns do not follow the regularities of equilocal distribution and no influence of arm lengths on band location is discernible. The following additional considerations and qualifications explain these exceptions.

1) *Transpositional heterochromatin amplification*. This offers a means for altering chromosomal size relationships and karyotype architecture. By accumulating large paracentromeric, intercalary and/or telomeric C-bands, the centromere-telomere relationships in the karyotype will change and subsequent rounds of telomeric heterochromatin amplification and/or transfer may then produce new interstitial C-band sites. *Leopoldia comosa* (see Bentzer & Landström 1975) is a striking example for the alteration of arm length by repeated addition of intercalary bands.
2) *Chromosome and karyotype mutations*. Any process of chromosomal rearrangement and karyotypic repatterning is likely to alter

the coordinates of a given C-band pattern. Bucci-Innocenti et al. (1983) provide an example where an intercalary C-band, which is equilocal relative to the telomere of the short arm of the chromosome has become dislocated by a pericentric inversion in *Triturus italicus*.

3) *Meiotic recombination*. Crossing-over between different structural morphs of C-banded chromosomes is also a cause of the diversification of C-band patterns (Kayano 1972).

4) *Methodological shortcomings and technical artifacts*. Some uncertainty has to be allowed in the position of C-bands as judged from mitotic metaphases. The proportion of euchromatin to heterochromatin segments at interphase may not be truly reflected in colchicine-condensed mitotic metaphase chromosomes.

   Moreover, interphase decondensation and hence the relative positions of telomeres to given intercalary sites are likely to vary, which means that the nuclear architecture determines only approximate regions (domains) for the initiation of intercalary bands rather than precise sites.

5) *Cases of independent chromosomal evolution*. A specific chromosome pair or a group of chromosomes may tend to exhibit a C-band pattern of their own. This phenomenon may be related to a special function of such chromosomes and/or to deviating chromosome mechanics and to their separated position within the nuclear architecture. Examples are: Nucleolus organizer chromosomes, sex chromosomes, chromosome 1 in crested and marbled newts (Sims et al. 1984) and the megameric chromosome pair in grasshoppers (John & King 1983).

   Holocentric chromosomes should be expected not to fit our model since they do not orientate according to the Rabl-polarisation.

6) *De novo origin of non-telomeric bands*. Our model can explain most of the so far described complex C-banding patterns on the assumption that repetitive sequences which stem from a telomeric site are dispersed to intercalary sites in a regular manner and that sequences can then spread further from these derived bands.

However, occasional *de novo* origin of C-bands by the spontaneous saltatory replication of any DNA sequence cannot be ruled out. A case for this may be the occurrence of a homogenously staining region, which shows all the characteristics of a C-band, in several wild mouse populations (Traut et al. 1984). Sometimes C-bands may originate at new intercalary sites as the remnants of inactivated centromeres in formerly dicentric chromosomes (see e.g. Therman et al. 1986).

CONCLUDING REMARKS

Our model implies that the non-random disposition of C-bands is not a result of selection processes operating on DNA sequences nor is it a consequence of chromosomal elements with specific functions related and/or unrelated to their particular localisations. The core of our model is therefore compatible with the assumption that heterochromatic DNA is 'selfish' or 'parasitic' (Orgel & Crick 1980; Doolittle & Sapienza 1980). Our mechanistic model contrasts with the concept of an optimal territory within the centromere-telomere field (Lima-de-Faria 1980) as far as heterochromatin is concerned. It has elements in common with a suggestion of Barnes et al. (1985) "that the localisation of particular repeat families to regions in a chromosome complement may be constrained simply by (mitotic) nuclear organisation". A similar conclusion reached by Schwarzacher et al. (1984) also considered restrictions exerted by the chromosomal disposition at meiosis.

In the model presented here, the effects of nuclear organisation of both mitosis and meiosis are analysed. The model assumes that the molecular mechanisms are restricted to specific stages of the nuclear division cycle. In its simplest form, the model suggests that the processes of heterochromatin amplification and transposition act predominantly in the mitotic interphase nucleus, whereas the mechanisms responsible for conservation of differences between C-bands and for the homogenisation of heterochromatic DNA sequences respectively act predominantly at some stages of first meiotic division.

Acknowledgements. We wish to thank Dr. G.P. Holmquist, Duarte, and Professor B. John, Canberra, for critical remarks on an earlier draft. The work was supported by grant No. P5080 of the Austrian Fonds zur Förderung der wissenschaftlichen Forschung.

REFERENCES

Ambros, P. 1983. PhD Thesis, Vienna.

Arnheim, N. 1983. pp 38-61 In 'Evolution of genes and proteins' eds M. Nei, R.K. Koehn, Sinnauer Associates, Inc., Sunderland.

Arnheim, N., Krystal, M., Schmickel, R., Wilson, G., Ryder, O. & Zimmer, E. 1980. Proc.Natl.Acad.Sci.USA **77**, 7323-7327.

Ashley, T. & Pocock, N. 1981. Genetica **55**, 161-169.

Barnes, S.R., James, A.M. & Jamieson, G. 1985. Chromosoma **92**, 185-192.

Bennett, M.D. 1982. pp 239-261 In 'Genome evolution' eds G.A. Dover, R.B. Flavell, Academic Press, New York.

Bentzer, B. & Landström, T. 1975. Hereditas **80**, 219-232.

Bird, A.P. 1980. Cell Biol. **3**, 61-111.

Blakey, D.H. & Vosa, C.G. 1981. Pl.Syst.Evol. **139**, 47-55.

Blakey, D.H. & Vosa, C.G. 1982. Pl.Syst.Evol. **139**, 163-178.

Brutlag, D.L. 1980. Annu.Rev.Genet. **14**, 121-144.

Bucci-Innocenti, S., Ragghianti, M. & Mancino, G. 1983. Chromosoma **88**, 208-215.

Charlesworth, B., Langley, C.H. & Stephan, W. 1986. Genetics **112**, 947-962.

Comings, D.E. 1980. Hum.Genet. **53**, 131-143.

Doolittle, W.F. & Sapienza, C. 1980. Nature **284**, 601-607.

Dover, G.A. & Flavell, R.B. 1984. Cell **38**, 622-623.

Flavell, R.B. 1982. pp 301-323 In 'Genome evolution' eds G.A. Dover, R.B. Flavell, Academic Press, New York.

Fussell, C.P. 1984. Genetica **62**, 193-201.

Greilhuber, J. 1984. pp 157-180 In 'Current concepts in plant taxonomy' eds V.H. Heywood, D.M. Moore, Academic Press, London, Orlando.

Greilhuber, J. & Loidl, J. 1983. p 344 In 'Kew chromosome conference II' eds P.E. Brandham, M.D. Bennett, George Allen & Unwin, London.

Greilhuber, J. & Speta, F. 1976. Pl.Syst.Evol. **126**, 149-188.

Greilhuber, J., Deumling, B. & Speta, F. 1981. Ber.Deutsch.Bot. Ges. **94**, 249-266.

Heitz, E. 1933. Planta **18**, 571-636.

Holmquist, G. & Dancis, B.M. 1980. Genetica **52/53**, 151-163.

John, B. & King, M. 1977. Chromosoma **65**, 59-79.

John, B. & King, M. 1980. Chromosoma **78**, 165-186.

John, B. & King, M. 1983. Chromosoma **88**, 57-68.

John, B., King, M., Schweizer, D. & Mendelak, M. 1985. Chromosoma **91**, 185-200.
Kayano, H. 1972. Japan.J.Genetics **47**, 219-221.
King, M. & John, B. 1980. Chromosoma **76**, 123-150.
Kurnit, D.M. 1979. Hum.Genet. **47**, 169-186.
Kurnit, D.M., Brown, F.L. & Maio, J.J. 1978. Cytogenet. Cell Genet. **21**, 145-167.
Lavania, U.C. & Sharma, A.K. 1985. J.Heredity **76**, 395-396.
Lima-de-Faria, A. 1980. Hereditas **93**, 1-46.
Loidl, J. 1983. Pl.Syst.Evol. **143**, 117-131.
Orgel, L.E. & Crick, F.H.C. 1980. Nature **284**, 604-607.
Rabl, C. 1885. Morphol.Jahrb. **10**, 214-330.
Schwarzacher, T. & Schweizer, D. 1982. Pl.Syst.Evol. **141**, 91-113.
Schwarzacher, T., Mayr, B. & Schweizer, D. 1984. Chromosoma **91**, 12-19.
Schweizer, D. & Ehrendorfer, F. 1976. Pl.Syst.Evol. **126**, 107-148.
Schweizer, D. & Ehrendorfer, F. 1983. Biol.Zbl. **102**, 637-655.
Sims, S.H., Macgregor, H.C., Pellatt, P.S. & Horner, H.A. 1984. Chromosoma **89**, 169-185.
Singer, M.F. 1982. Int.Rev.Cytol. **76**, 67-112.
Smith, G.P. 1976. Science **191**, 528-535.
Therman, E., Trunca, C., Kuhn, E.M. & Sarto, G. 1986. Hum. Genet. **72**, 191-195.
Traut, W., Winking, H. & Adolph, S. 1984. Cytogenet.Cell Genet. **38**, 290-297.
Vosa, C.G. 1979. Pl.Syst.Evol. **132**, 141-148.
Wilson, E.B. 1925. The Cell in Development and Heredity. 3rd Ed. Macmillan, New York.
Young, B.S., Pession, A., Traverse, K.L., French, C. & Pardue, M.L. 1983. Cell **34**, 85-94.

The role of chromosomes in speciation: a new interpretation

B. Dutrillaux[1] and Y. Rumpler[2]

[1]CNRS U.A. 620, Institut Curie, Section de Biologie, 26 rue d'Ulm, 75231 Paris Cédex 05, France, and [2]Institut d'Embryologie, Strasbourg, France

Whatever the importance given to chromosome rearrangements during evolution, almost all the interest was concentrated onto their short term consequences, and particularly their role of gametic barrier, due to the reduced fertility of heterozygotes carriers (Bengtsson, 1980, Fredga, 1977, Dutrillaux, 1975, Lande, 1979, Lejeune, 1975, White, 1973, Wilson et al., 1975).

We would like to point out that chromosomal rearrangements may play a larger role, by long term effects on the modes of speciation. They may have major consequences, on genic mixing, changing drastically the effect of endogamy.

A. <u>The changes of chromosome numbers</u>

The fact that chromosome numbers (2N) change from species to species is very well known. A nice synthetic paper was published by Matthey (1973), showing that in mammals, for instance, 2N range from 6 to 84, with a clear mode between 38 and 48. In a given order, like that of Primates, 2N is comprised between 20 and 80. However, the distributions of 2N vary from one group (infra- or sub-order) to another : asymptotical in

Chromosomes today, volume 9 Allen & Unwin 1987

Catarrhini, Gaussian in Platyrrhini and flat in Prosimii (Dutrillaux et al., 1984). Indeed, the variation of chromosome numbers is determined by the types of rearrangements having preponderantly occurred during the evolution of such or such group. For instance, Robertsonian translocations progressively reduced 2N in Lemuridae (Rumpler and Dutrillaux, 1976, 1979) whereas fissions increased 2N up to 72 in Cercopithecinae (Dutrillaux et al., 1982).

This tendency for an accumulation of a given type of chromosomal rearrangement was previously noticed in other animals, like insects, which was interpreted as the result of selection : orthoselection (White, 1973). In our opinion, it could also result from specificities of mutagenesis (Dutrillaux et al., 1981) and the term of orthomutation might have been proposed as well.

Whatever its cause, this tendency should have important consequences on the gametic variability, which obviously depends on chromosome numbers. For the moment, we shall neglect the occurrence of crossing-over which will be considered after. In an individual of a given species, the number of possible different gametes, due to the random distribution of paternal and maternal chromosomes of each pair, is given by 2 , where N = haploid number of chromosomes.

This elementary fact may have consequences for the reproduction of populations. For instance, in Cercopithecidae, all Papioninae (baboons and macacas) possess 21 chromosome pairs, whereas some Cercopithecinae, like C. mitis have 36. Thus, the possibilities of meiotic segregations are 2 = 32,768 higher in the latter than in the former. In Prosimii, where Tarsius possesses 80 and Lepilemur ruficaudatus 20 chromosomes, the difference of the number of possible segregations in higher than one thousand million.

Obviously, the parental combinations of genes will be inasmuch rapidly disrupted that the number of possible segregations will be high, indicating that the genetic effect of endogamy will be related to chromosome number, for a part.

B. The modification of genetic distance between linked genes.

The effect of chromosome number on meiotic segregation could be largely balanced by gene recombinations. However, the occurrence of recombinations is itself largely dependant on the karyotype of the species considered, and thus on the chromosome rearrangements which formed it.

As a matter of fact, some of the consequences of chromosomal rearrangements have been noticed a long time ago. For instance, a Robertsonian translocation creates a linkage between the 2 groups of genes close to the centromere which were totally independant before. On the contrary, fissions suppress existing linkages, and inversions both suppress linkages and created others.

It was postulated that the rearrangements could lead to the formation of favourable combinations, giving a selective advantage to the carrier by comparison to its population of origin.

Our purpose is not to minimize this possible role of the chromosomal rearrangements, but we think that another consequence may be more important. It consists in changing the ratio between physical and genetical distances of the genes. In other words, we would like to show that 2 genes, whose respective position did not change may have acquiered quite different recombination rate.

1) Chiasmata give a valuable indication of crossing-over

Many ideas on chiasma movement and terminalization became obsolete when it could be demonstrated, by BrdU incorporation experiments, that chiasmata are observed at the place where crossing-over took place (Tease, 1978, Allen, 1979, Jones and Tease, 1979, Polani et al., 1979, Kanda and Kato, 1980). It could be shown also that the genetic distances estimated either by chiasma count or by gene recombination were compatible, for a given chromosome (Hulten et al., 1982, Human Gene Mapping, 1985).

Thus, we shall consider that the studies on chiasma counts give valuable estimates of gene recombination, in a given chromosome or karyotype.

2) The chiasma distribution is characteristic of a given chromosome.

Classical studies on Drosophila have shown that, for a given chromosome, there is not a strict relationship between the physical distance between 2 genes and their recombination rates, those close to the centromere being less frequently recombined than distal ones. (Haldane, 1931, Mather, 1938).

More recent studies on human meiosis have brought the same conclusions. For instance, on chromosomes 1, 2 and 9, there are much more chiasmata at terminal or subterminal than at proximal segments (Laurie and Hulten, 1981).

If these results have a general value, it means that the genetic distance between 2 genes may vary according to their position on a given chromosome, even if no rearrangements occurred between their 2 locus.

3) The relative number of chiasmata depends on chromosome length and morphology i.e., on chromosome arms

For reasons which seem related to meiotic segregation, their is one chiasma per bivalent at least, whatever the number of chromosomes of the karyotype considered. Thus, the same minimal limitation existing for the smallest as for the largest, more chiasmata per length unit can be expected on small than on large bivalents.

At figure 1, we have represented the relationship between the relative number of chiasmata and the relative length of chromosomes in human male meiosis (calculated from Laurie and Hulten, 1985). It clearly shows that the number of chiasmata, per length unit, is inversly proportional to chromosome length.

At figure 2, we have made the same representation for chromosome arms. For them also, the shorter they are, the more they have chiasmata in proportion to their length. The euchromatic arms of acrocentrics tend to have relatively more chiasmata than those of non acrocentrics although the acrocentrics tend to have less chiasmata than non acrocentrics (fig. 1).

This appearant contradiction is due to the fact that the acrocentrics carry more heterochromatin, in which there are no chiasmata.

4) The number of chiasmata per diakinesis, differing from species to species, is determined by chromosome numbers and morphologies.

If the observations made on human chromosomes have a general value, it should be possible to predict the number of chiasmata at diakinesis, by the characteristics of the mitotic karyotype of a given species : it should be related to the number and the morphology of the chromosomes.

In particular, the number of chiasmata should be determined by the number of chromosome arms of the haploid complement i.e. by half Nombre Fondamental : $\frac{NF}{2}$. Thus a karyotype with n acrocentrics should have less chiasmata than a karyotype with n metacentrics or submetacentrics. In spite of the increase of their relative size, a decrease of the chromosome numbers

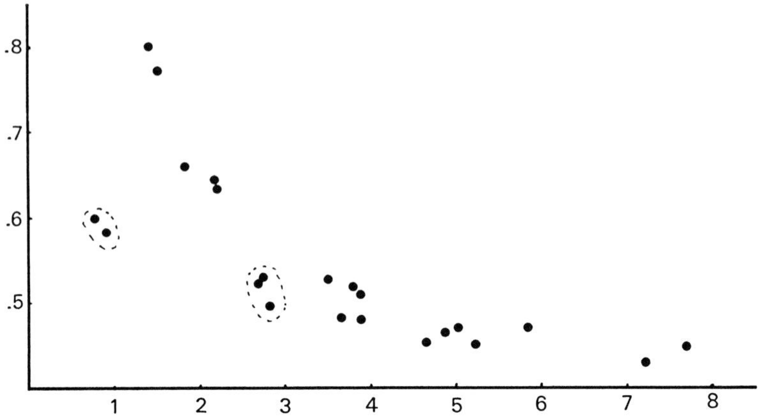

Fig. 1 Relationships between the relative number of chiasmata (ordinate) (mean number of chiasmata/relative length of the chromosome) and the relative length of the 22 human male autosomal bivalents (abscissa). Calculated from Laurie and Hulten, 1985. Acrocentrics 13 to 15 and 21-22 are surrounded with dotted line.

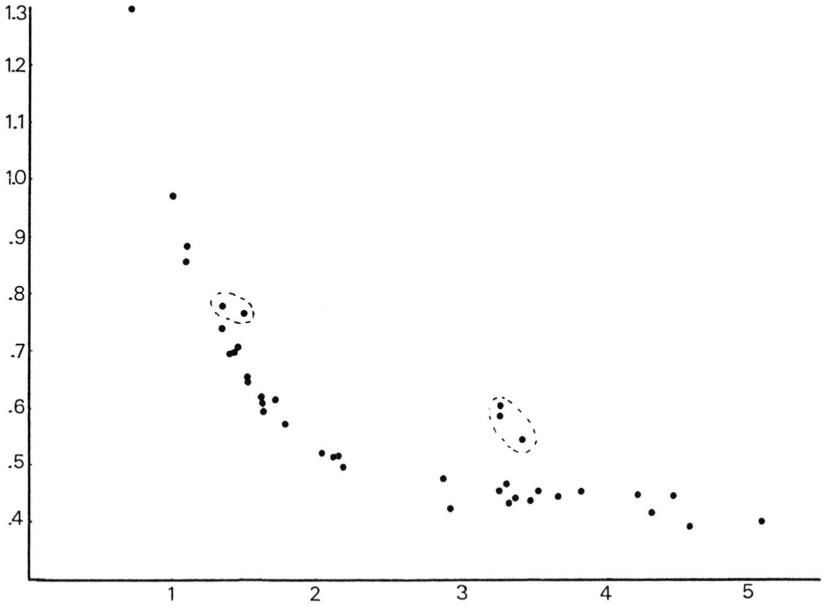

Fig. 2 Relationships between the relative number of chiasmata (ordinate) and the relative length of the 39 euchromatic arms of the human male autosomal bivalents (abscissa). Calculated from Laurie and Hulten, 1985. Long arms from acrocentrics 13 to 15 and 21-22 are surrounded by a dotted line.

should be accompanied by a decrease of the number of chiasmata, except in case of Robertsonian evolution which do not change the $\frac{NF}{2}$.

Some examples of Chiasma counts (NC) in Primates are given in table I, which also indicates the ratio (coefficient a) of $\frac{NC}{NF/2}$. Coefficient a is comprised between 1.10 and 1.25 and has an average values of 1.15.

In Rodentia (table I), NC is also generally higher than NF except in 3 cases. Coefficient a varies from .86 to 1.87, and its average value is 1.17. These data are represented at fig. 3, which shows that a grossly linear relationship exists, giving a curve of the type $y = ax + b$, where $y = NC$, $a = .77$, $x = \frac{NF}{2}$ and $b = 8.26$.

From table I and figure 3, it can be expected than the relationship between NC and NF is not linear for the extreme values of $\frac{NF}{2}$. The meiotic study of more animals with very low or very high chromosome numbers and NF should be informative in this respect.

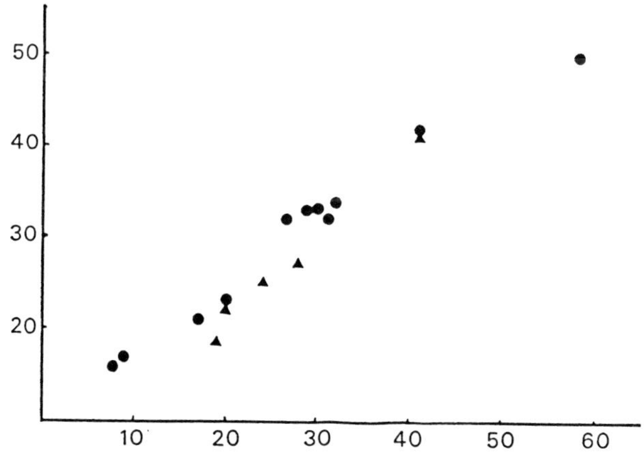

Fig. 3 Relatioships between the total number of chiasmata (ordinate) and the half Nombre Fondamental (abscissa) in 16 rodent species. Circles correspond to reference 3, and triangles to references 2, 5 and 6 of table I.

In spite of the scarcity of the data considered here, it is clear that the average number of chiamata varies from species to species. The range of this variation may be higher than a factor 3 in mammals.

5) The amount of gene recombinations should vary from karyotype to karyotype

By the comparative study of about 100 Primate species, of Lagomorpha, Rodentia and Carnivora, it could be largely demonstrated that a very good conservation of euchromatic bands existed during evolution (Dutrillaux, 1979, Dutrillaux et al., 1980, 1981, Dutrillaux and Couturier, 1983 and Petit et al., 1984).

This conservation was greatly confirmed by comparative gene mapping which showed that the chromosomes considered as homoelogous, by banding comparisons, also carry the same genes (Human Gene Mapping, 1985).

Thus, even when the chromosome numbers and morphologies vary drastically, the genetic material is preserved.

All major quantitative variations of chromosomal material are related to variation of heterochromatic segments. When large heterochromatic segments exist, like in hedgehog (Natarajan and Gropp, 1971) or in Capucin monkey (Dutrillaux, 1979), no chiasmata affect them.

Thus, it can be deduced that heterochromatin addition or deletion should not drastically change the number of chiasmata per cell.

Gene mapping has also largely shown that the genes are localized on all chromosome segments, except heterochromatin.

Finally, if chiasmata really do represent directly crossing-over, we must conclude that genic recombination rates, in general, largely vary from species to species. It could be of a factor 3 in

Species	N	$\frac{NF}{2}$	N C	Coefficient a	ref
Primates					
Human	23	40	50 ± 2.1	1.25	1
Mandrillus sphinx	21	41	45 ± 2.8	1.10	2
Cebus capucinus	27	35	41 ± 1.7	1.17	2
Lemur fulvus	30	32	36.5 ± 2.3	1.15	2
L.f. collaris	26	32	35.5 ± 1.3	1.11	2
L.f. albocollaris	24	32	36 ± 2.7	1.12	2
Rodentia					
Akodon arviculoides	7	9	16	1.67	3,4
Akodon arviculoides	7	8	15	1.87	3,4
Akodon sp	12	20	23	1.15	3
Akodon nigrita	26	26.5	32	1.21	3
Zygodontomys lasiurus	17	17	21	1.24	3
Clyomys laticeps	17	30	33	1.10	3
Nectomys squamipes	28	29	33	1.14	3
Nectomys squamipes	29	31	32	1.03	3
Oxymicterus sp	27	32	34	1.06	3
Euryzygomatomys guiara	23	41	42	1.02	3
Proechymys iheringi	31	58	50	.86	3
Mus musculus	20	20	24	1.20	5
Mesocricetus auratus	22	28	26.5	.95	6
Cricetulus griseus	11	19	17	.90	6
Gerbillus aureus	36	40	41 ± 1.3	1.02	2
Taterillus gracillis	18	23	25 ± .8	1.09	2

Table I Haploid chromosome number (N), half number fondamental (N F), number of chiasmata (N C), coefficient a : (N C / $\frac{NF}{2}$, in various species of Primates and Rodentia. References 1 : Laurie and Hulten (1985) and ISCN (1977), 2 : personnal data, 3 : Yonenaga (1972), 4 : Yonenaga - Yassuda (1979), 5 : Polani and Jagiello (1976), 6 : Fredga and Santesson (1964). For references 3 and 6, chiasma counts not given by the authors, were estimated by ourselves, on the published figures. Thus these counts are approximative, and no standard deviations are given. For reference 5, we have retained an average value of those published.

Rodentia, for instance.

C. Changes of chromosomes and of genetic recombination rates are synergic, and many influence the mode of speciation.

In karyotypes with small chromosome numbers, the low number of possible meiotic segregations is associated with a low recombination rate per length unit of the genome. Both restrict the genetic mixing.

In karyotypes with large chromosome numbers, the high number of possible meiotic segregations is associated with a high recombination rate. Both increase the genetic mixing.

Indeed, the genetic mixing may have long term consequences on population genetics. The higher it will be, the faster groups of mutant alleles will be dissociated. This will minimize the effects of endogamy. In the very probable hypothesis that a single mutation is not sufficient to lead to speciation, we can assume that a simple dichotomic speciation will have a low probability to occur if the genetic mixing is high.

Can we predict the consequences of the various modes of chromosome evolution ?

A Robertsonian evolution will reduce 2N, at most by a factor 2 if it started on a karyotype composed of acrocentrics only, thus, it may reduce the possible chromosome segregations of $2^{N/2}$ (N : haploid number of chromosomes of the ancestral karyotype) at most. It will make new linkage groups for genes near to the centromeres, and it will slightly reduce the recombination rates for genes located on the same arms.

An evolution by end-to-end fusions, will have the same consequences, but it may be more efficient. Firstly, their is less limitation to the reduction of chromosome number. Secondly, the NF will not be conserved. The reduction of the number of arms will be of course associated by their increase in size, and thus a decrease of recombination rate per length unit. This decrease may be of a factor 3, as we have seen above.

An evolution by fissions will have just opposite consequences : a considerable increase of both chromosome segregations and of genic recombinations.

An evolution by inversions will play a role by modifying the position of the genes only, and thus eventually their recombination rates. Thus its long term efficiency may be not very high, as far as genetic mixing is concerned.

If the above considerations are correct, a reduction of N and NF should be observed in the groups which have undergone a dichotomic speciation, whereas higher values of N and NF should be observed in others. Unfortunately, the phylogeny of a very small number of mammals was studied in detail, although their is a large amount of publications on this question. The reserves that we have about published data are due to the following difficulties of interpretation that we met ourselves : when few species of a group are considered, a dichotomic evolution appears likely, but when more species are studied, it becomes obvious that the chromosomal rearrangements are diversly shared by several of them. Then, the dichotomic model must be rejected.

In Primates, where we have studied about half living species, we have now an accurate information for most Catarrhini and Prosimii (Dutrillaux et al., 1985). Obviously, pure dichotomic examples of speciation appear quite rare, the only clear one being that of Lepilemuridae (Rumpler et al., 1985, in preparation). In this family, where N = 13.5 on the average, many end-to-end translocations occurred, reducing NF/2 to 21. Two other examples in Prosimii are less demonstrative, since they mostly correspond to "chromosomal sub speciation", in the species Lemur fulvus and Hapalemur griseus, for which N = 27 and NF/2 = 32.

On the contrary, many examples of non dichotomic speciation exist in Primates. The most extreme one, in which 10 chromosomal rearrangements are shared by a dozen species or subspecies is that of a group of Cercopithecus : C. mona, C. cephus, C. ascanius, C. nictitarus, C. hamlyni and C. neglectus, for which N 34 and NF/2 58 on the average. These values are among the highest in Mammals.

In Rodentia, we have studied Gerbillidae in some detail (Benazzou et al., 1984, Viegas-Péquignot et al., 1985). They may be separated into 3 inequal groups. One large, composed of Meriones, Psamomys, Pachyuromys, Tatera and Gerbillurus species. They have undergone a non dichotomic evolution. Their average values of N is 25 and NF/2 : 36. In the second group, composed of Gerbillus species, it was not possible to propose an evolutive scheme. In the third and small group, composed of Taterillus species, a dichotomic scheme was reconstructed : N = 18 and NF/2 = 23 on the average.

In Carnivora, the chromosomal phylogeny of several families was reconstructed (Couturier and Dutrillaux, 1985). For Felidae, Procyonidae and Viveridae, a clearly not dichotomic evolution was found. Their average values of N are 19,19 and 20,5, and of NF/2 : 35, 34 and 33 respectively. Pinnipedia was the single dichotomic group : N = 17, NF/2 = 32. Their chromosome numbers are slightly lower than those of the other carnivora. In Mustelidae, a dichotomic evolution was found for some species, but the number of species studied was too low for being really representative.

Finally, although dichotomic models of speciation are widely accepted, they seem to have rarely occurred, at least in the 3 orders of mammals that we have partially studied. In most cases, chromosomal rearrangements are shared in diverse combinations by several species, and thus a populationnal evolution must took place.

Its seems that karyotypes with large chromosome

numbers and high recombination rates may prevent dichotomic evolution, by suppressing in a few generation the linkage of most groups of mutant alleles that might play a role in speciation.

On the opposite, karyotypes with low chromosome numbers and low recombination rates will recombine less frequently the groups of mutant genes, and the occurrence of groups of new characters will be less improbable.

The fact that most models of evolution, which gave a large part to dichotomy, were elaborated from the study of organisms with low chromosome number, like Drosophila may not be without signification.

SUMMARY : The number of possible meiotic segregations of paternal and maternal chromosomes is 2^N (N = haploid chromosome number). In Eutherian Mammals, where large variations of N exist, 2^N may be very different in related species, genera or families. For instance, in Cercopithecidae, species with the highest values (N = 36) make 2^{15} = 32,768 more gametic chromosome combinations than those with the lowest values (N = 21). It is also shown that the number of chiasmata varies from species to species and is proportionnal to the haploid number of arms (NF/2) of the karyotype. Thus, increase or decrease of both N and NF/2, often concomitant, both increase or decrease genetic mixing. The study of chromosomal phylogeny of Primates, Carnivora and Rodentia shows that pure dichotomic evolution is rare, at contrast with populational (network) evolution which seems most frequent. In these Mammals, the few examples of dichotomic evolution are observed in groups with low values of N and NF/2. It is concluded that dichotomic evolution, which should be favoured by the transmission of groups of linked mutant alleles, is prevented, in most instances, by high recombination rates in karyotypes with high values of N and NF/2. Thus, the mode of speciation would depend on

Karyotype modification by long term effects on population genetics, in addition to the immediate effects of gametic barrier due to chromosomal rearrangements at the heterozygote status.

REFERENCES

Allen, J.W. 1982. BrdU-dye characterization of late replication and meiotic recombination in Armenian Hamster germ cells. Chromosoma. 74, 534-543.

Benazzou, T., E., Viegas-Péquignot, M., Prod'homme, M., Lombard, F., Petter and B. Dutrillaux 1984. Phylogénie chromosomique des Gerbillidae. III. Etude d'espèces des genres *Tatera*, *Taterillus*, *Psammomys* et *Pachyuromys*. Ann. Génét. 27, 17-26.

Bengtsson, B. 1980. Rates of karyotypes evolution in placental mammals. Hereditas. 92, 37-47.

Couturier, J. and B. Dutrillaux 1986. Evolution chromosomique des carnivores. Mammalia. 50, 124-163.

Dutrillaux, B. 1975. Sur la nature et l'origine des chromosomes humains. L'Expansion Scientifique.

Dutrillaux, B. 1979. Chromosomal evolution in primates : tentative phylogeny from *Microcebus murinus* (prosimian) to man. Hum. Genet. 48, 251-314.

Dutrillaux, B., E., Viegas-Péquignot and J. Couturier 1980. Très grande analogie de marquage chromosomique entre le lapin (*Oryctolagus cuniculus*) et les primates, dont l'homme. Ann. Génét. 23, 22-25.

Dutrillaux, B. and J. Couturier 1983. The ancestral karyotype of Carnivora : comparison with that of platyrrhine monkeys. Cytogenet. Cell Genet. 35, 200-208.

Dutrillaux, B., J., Couturier and E. Viegas-Péquignot 1981. Chromosomal evolution in primates. Chromosomes today 7. 176-191.

Dutrillaux, B., J., Couturier, M., Muleris, M. Lombard and G. Chauvier 1982. Chromosomal phylogeny of forty-two species or subspecies of cercopithecoids (Primates Catarrhini). Ann. Génét. 25, 96-109.

Dutrillaux, B., M. Muleris and M. Paravatou-Petsota 1984. Diagrammatic representation for chromosomal mutagenesis studies. I. Karyotypes most similar to that of man. Mutation Res. 126, 81-92.

Dutrillaux, B., J., Couturier, M., Muleris, Y., Rumpler and E. Viegas-Péquignot 1986. Relations chromosomiques entre sous-ordres et infra-ordres, et schéma évolutif général des Primates. Mammalia. 50, 108-123.

Fredga, K. and B. Santesson 1964. Male meiosis in the syrian, chinese, and european hamster. Hereditas. 52, 36-48.

Fredga, K. 1977. Chromosomal changes in vertebrate evolution. Proc. R. Soc. Lond. B. 199, 377-397.

Haldane, J.B.C. 1931. The cytological basis of interference. Cytologia. 3, 54-65.

Hulten, M.A., R.W., Palmer and D.A. Laurie 1982. Chiasma derived genetic maps and recombination fractions : chromosome 1. Ann. Hum. Genet. 46, 167-175.

Human Gene Mapping VIII. 1985. Helsinki Conference 1985. Eight international workshop on human gene mapping. Cytogenet Cell Genet. 40.

ISCN. 1978. An international system for human cytogenetic nomenclature. Birth defects : original article series, vol. 14, N° 8 (The national Foundation, New York 1978) ; also in Cytogenet. Cell Genet. 21, 309-404.

Jones, G.H. and C. Tease 1979. Analysis of exchanges in differentially stained meiotic chromosomes of Locusta migratoria after BrdU-substitution and FPG-staining. III. A test for chromatid interference. Chromosoma. 73, 85-91.

Kanda, N. and H. Kato 1980. Analysis of crossing-over in mouse meiotic cells by BrdU labelling technique. Chromosoma. 78, 113-121.

Laurie, D.A., M.A. Hulten and G.H. Jones 1981. Chiasma frequency and distribution in a sample of human males : chromosomes 1, 2 and 9. Cytogenet. Cell Genet. 31, 153-166.

Laurie, D.A. and M.A. Hulten 1985. Further studies on bivalent chiasma frequency in human males with normal karyotypes. Ann. Hum. Genet. 49, 189-201.

Lande, R. 1979. Effective deme sizes during long-term evolution estimated from rates of chromosomal rearrangement. Evolution. 33, 234-251.

Lejeune, J. 1975. Sur les mécanismes de la spéciation. C.R. Soc. Biol. 169, 828-844.

Mather, K. 1938. Crossing-over. Biol. Rev. 13, 252-292.

Matthey, R. 1973. The chromosome formulae of eutherian mammals, in : A.B. Chiarelli and E. Capanna (Eds.), Cytotaxonomy and Vertebrate Evolution. Academic Press, New York, pp. 531-616.

Natarajan, A.T. and A. Gropp 1971. The meiotic behaviour of autosomal heterochromatic segments in hedgehogs. Chromosoma, Berl. 35, 143-152.

Petit, D., J., Couturier, E., Viegas-Péquignot, M. Lombard and B. Dutrillaux 1984. Très grande similitude entre le caryotype ancestral des écureuils (rongeurs) et celui des primates et des carnivores. Ann. Génét. 27, 201-212.

Polani, P.E., J.A., Crolla, M.J. Seller and F. Moir 1979. Meiotic crossing-over exchange in the female mouse visualised by BUdR substitution. Nature. Lond. 278, 348-349.

Polani, P.E. and G.M. Jagiello 1976. Chiasmata, meiotic univalents, and age in relation to aneuploid imbalance in mice. Cytogenet. Cell Genet. 16, 505-529.

Rumpler, Y. and B. Dutrillaux 1976. Chromosomal evolution in Malagasy lemurs. I. Chromosome banding studies in the genuses Lemur and Microcebus. Cytogenet. Cell Genet. 17, 268-281.

Rumpler, Y. and B. Dutrillaux 1979. Chromosomal evolution in Malagasy lemurs. IV. Chromosome banding studies in the genuses Phaner, Varecia, Lemur, Microcebus and Cheirogaleus. Cytogenet. Cell Genet. 24, 224-232.

Rumpler, Y., B., Ishak, S., Warter and B. Dutrillaux 1985. Chromosomal evolution in Malagasy Lemurs. VIII Chromosome banding studies of Lepilemur ruficaudatus, L. Leucopus and L. septentrionalis. Cytogenet. Cell Genet. 39, 194-199.

Tease, C. 1978. Cytological detection of crossing-over in BUdR substituted meiotic chromosomes using the fluorescent plus Giemsa technique. Nature. Lond. 272, 823-824.

White, M.J.D. 1973. Animal cytology an evolution. Cambridge University press.

Wilson, A.C., G.L., Bush, S.M., Case and M.C. King 1975. Social structuring of mammalian populations and rate of chromosomal evolution. Proc. nat. Acad. Sci. U.S.A. 72, 5061-5065.

Yonenaga, Y. 1972. Polimorfismos chromosomicos em roedores brasilieros. Sao Paulo. Thesis.

Yonenaga-Yassuda, Y. 1979. New karyotypes and somatic and germ-cell banding in Akodon Arviculoides (Rodentia, Cricetidae). Cytogenet. Cell Genet. 23, 241-249.

Chromosome evolution in marsupials

D. L. Hayman, R. H. Rofe and P. J. Sharp

Department of Genetics, University of Adelaide, Australia

The development of chromosome banding techniques has resulted in considerable knowledge of the comparative cytogenetics of many groups of mammals. In order to assess the significance of the results obtained from studying marsupial cytogenetics it is necessary to indicate the special features of this group of mammals.

Marsupials are one of the three living lines of mammalian evolution. Comparative studies of the amino acid differences between the eutherian and marsupial haemoglobins suggest that these two groups diverged some one hundred and thirty million years ago (Air et al., 1971). The oldest fossil remains of marsupials are found in North America. The distribution of living species shows some 77 species to be found in Central and Southern America and some 161 species in Australia and some of the islands to the north of Australia (Kirsch and Calaby, 1977).

It seems most likely that the initials of the Australian marsupial fauna reached Australia through an Antarctic connection which disappeared about fifty million years ago (Keast, 1977; Woodburne and Zinsmeirer, 1983). Most significantly this migrant fauna contained no representatives of eutherian mammals; thus the only mammals present in Australia until comparatively recent times were marsupials. The Australian marsupials evolved free from eutherian competition to produce a tremendous diversity of species adapted to many different environments. In South America marsupial evolution occurred with the competitive presence of eutherian species, and a less diverse fauna is found. Consequently marsupials are not only a group with which to compare eutherian mammals; this ancient dichotomy between these present day marsupial faunas enables an internal comparison to be made of the chromosome evolution between the two geographically isolated groups.

TABLE 1 Chromosome numbers in marsupials.

Superfamily	Family	Number of species	2n (frequency)
Didelphoidea	Didelphidae	70	*14(12), 18(4), 22(7)
	Microbiotheriidae	1	*14(1)
Caenolestoidea	Caenolestidae	7	*14(4)
Dasyuroidea	Dasyuridae	56	*14(44)
	Myrmecobiidae	1	*14(1)
Perameloidea	Peramelidae	16	*14(8)
	Thylacomidae	2	18♀ 19♂(1)
Phalangeroidea	Phalangeridae	11	*14(2), 20(3)
	Burramyidae	7	*14(7)
	Petauridae	22	10(1), 12(1), 16(4), 18(3), 20(8), 22(5)
	Macropodidae	56	10♀ 11♂(1), 12(2), 12♀ 13♂(1), 14(3), 16♀ 15♂(1), 16(11), 18(3), 20(3), 22(13), 24(1), 32(1)
Vombatoidea	Vombatidae	3	*14(2)
	Phascolarctidae	1	16(1)
Tarsipoidea	Tarsipedidae	1	24(1)
Notoryctoidea	Notoryctidae	1	20(1)

2n = *14 conserved complement

The breadth of marsupial adaptations is reflected in their taxonomic treatment. Some authorities have suggested that there are several orders of marsupials and most recognise the presence of superfamilies (Kirsch and Calaby, 1977). Some indication of the age and divergence of the different superfamilies comes from the serological studies of Kirsch (1977) who showed that the representatives of the two American superfamilies Didelphoidea and Caenolestoidea) are no more like each other than they are like representatives of the Australian superfamilies.

The general details of chromosome number and morphology are known for more than half the species of Australian marsupials and for a representative sample of the American species (Table 1). Marsupials show the following general cytological features:-

(1) Low diploid chromosome numbers ranging from 2n=10 to 2n=32. The low chromosome number complements have large chromosomes.
(2) Nuclear DNA contents range from values slightly less than Man to values some 40% more.
(3) There are no substantial differences between the chromosome complements of the two separate marsupial faunas with respect to number and morphology. This suggests that the different selective histories of the two faunas have been without substantial effect on these characters.
(4) The commonest chromosome complement found in both faunas is 2n=14, with three large metacentric or submetacentric chromosomes, one medium sized near metacentric pair, two pairs of smaller chromosomes and the sex chromosome pair (Fig. 1).

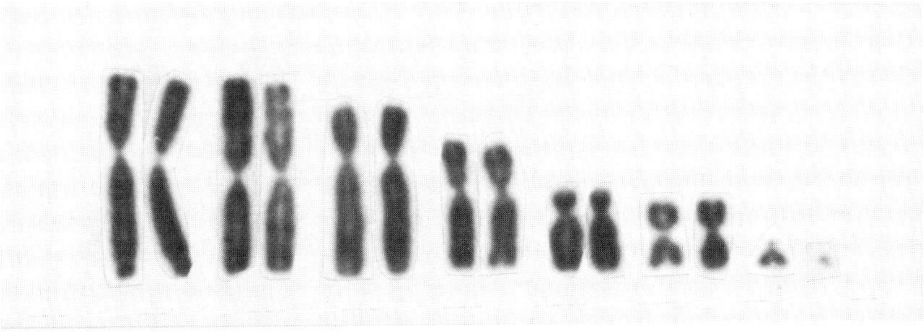

FIGURE 1 A chromosome complement representative of the postulated 2n=14 conserved complement.

This form of complement is present in many of the superfamilies of both faunas. It is the only complement observed in the Australian Dasyuroidae (in 44 species studied out of 56 identified) and in the American Caenolestoidae (4 species out of 7). The widespread distribution of this form of complement suggested that it may represent a conserved, possibly ancestral complement (Hayman and Martin, 1969, 1974; Hayman et al., 1971). The slight variations in chromosome morphology which may occur between species were argued to be the result of changes internal to the chromosome resulting from inversions or internal transpositions.

The argument that the 2n=14 complement is a conserved complement means that those complements greater than 2n=14 must have been derived from it by a process of chromosome fission, i.e. an increase in both chromosome and centromere number. Since there are many such complements in the different superfamilies, chromosome fission must have occurred frequently.

An alternative view of marsupial chromosome evolution is to argue that the ancestral chromosome complement is one greater than 2n=14, such as 2n=22, since this is found in many marsupial genera including Didelphis which has some similarities to the features of early marsupial fossils (Matthey, 1973; Sharman, 1973, 1974). Were this or another high chromosome number to be the ancestral complement, then the major evolutionary mechanism would be one resulting in reductions in chromosome number and centromere numbers, i.e. chromosome fusion, a well established mechanism in the evolution of other groups of organisms. It has the advantage of involving mechanisms which are easier to conceive than are those of fission.

These two contrasting hypotheses are based upon chromosome number and morphology alone. It follows from the argument that the 2n=14 complement is a common conserved complement that homology should be present at the level of G-banding. Such an analysis has now been completed for a broad range of Australian species and one American species with 2n=14 (Rofe and Hayman, 1985). The G-banding patterns of each of the individual chromosomes of the different chromosome complements show remarkably close agreement. The differences in gross morphology of the individual chromosomes which exist between different species can be shown to be the result of intra-chromosomal changes. Frequently the variations in morphology are common to the chromosomes of species from different superfamilies suggesting that the same structural variant has been selected in different lineages. Particularly critical evidence for the similarity is shown by the G-banding pattern of the single American species available for study. This encourages the view that when further data from American species are available it should be possible to propose a highly detailed G-banding complement which may be ancestral to both groups.

The possession of similar G-banding chromosomes in the complements of these diverse species restricts the fusion hypothesis. If the present day marsupial complements evolved from an ancestral marsupial complement with more than 2n=14, then the same fusions must have occurred independently in each marsupial lineage. The small changes in morphology observed between some of the different species are then the result of changes after any fusion events. It seems an inescapable conclusion that a common 2n=14 G-banding complement must have been present in the different lineages early in their evolution and is retained in many of them today.

The fission hypothesis is greatly strengthened by these G-band observations. Had there been evidence of the possessions of different combinations of G-band homologous arms by different species, this could have argued for fusion rather than fission. Further support comes from the studies of G-banding in the large marsupial family Dasyuroidae, comprising some fifty-six species, twenty-six of which have now been G-banded. No substantial difference in the G-banding pattern between these species has been detected (Young et al., 1982; Westerman pers. commun.).

Increases in chromosome number require increases in centromere and telomere number. On classical Mullerian grounds such increases must involve a dispensable donor chromosome and a complex series of exchanges (Muller, 1940). However while the mechanisms proposed by Muller are perfectly feasible they should not be considered as exclusive. Telomeres have been shown to arise following healing of broken chromosome ends (McClintock, 1941) and centromeres may also arise from latent centromeres. Clearly chromosome evolution must involve increases in chromosome number as well as the more readily understood decreases in number. The reluctance to accept increases in chromosome number has been largely due to the apparent difficulties posed by the mechanics.

The marsupials provide a remarkable example of evolutionary diversification in morphology, physiology and behaviour with a minimum of changes in the phenotype of the conserved complement of many species. There are some changes in the sex chromosomes which differ substantially in size between some species, usually by the addition of C-banding material. The nucleolar organiser region has no fixed location. It may be present either on autosomes or the sex chromosomes and in some species is present on both. Whether this mobility is a feature of multigene families alone is not known.

There is little evidence in the marsupials to support the argument that chromosomal change accompanies or causes speciation. The major theme in marsupial evolution is long term conservation of chromosomal and presumably linkage group complements. It is difficult to escape the conclusion that there are substantial constraints on the type of change that can occur, or can be incorporated into this marsupial karyotype. Yet these constraints have not been an impediment to marsupial evolution - the conserved complement has served the marsupials well.

Since there has been substantial genetic divergence between species with this conserved complement, it seems that there are two levels of variation upon which selection is operating, one is at the level of the chromosome complement and the other at the level of the genetic complement. The idea of such a division is not novel - a different form of example is shown by the genus Plethodon (Macgregor, 1982). What is not yet understood is the nature of the constraints operating to restrict changes to the karyotype.

In contrast to the general pattern of constancy of chromosome complement, there are marsupial families in which the species show considerable diversity in chromosome number and morphology. This is particularly evident in the Australian Phalangeroidea which has the widest range of chromosome numbers. G-banding analyses of a number of these presumably derived complements show that all or a substantial proportion of these chromosome complements can be related simply to the 2n=14 complement.

The chromosomes of the koala (Phascolarctus cinereus) 2n=16, the only known species of the Vombatoidea which does not have the conserved complement, may be derived from this complement by centric fission of one chromosome, otherwise the autosomal homology is very close. The short arms of one of these derived chromosomes, if it exists at all, is very small.

The autosomal chromosomes of Tarsipes rostratus (2n=24) are all acrocentric or telocentric (Hayman and Sharp, 1982). They can be derived from the conserved complement by four centric fissions and one other fission. The G band correspondence is almost complete.

The autosomal chromosomes of Trichosurus vulpecula (2n=20) (Phalangeroidae) can be derived by one non-centric and three centric fissions (Rofe, 1979). Again, the G-band homology is very close and little differentiation has occurred.

Less complete homology is found with the chromosomes of a Petaurid (Petaurus norfolkensis 2n=22) and the macropod Thylogale billardieri (2n=22) (Rofe, 1979). Again there is evidence of both centric and non-centric fissions and only some 10% of the haploid complement of these two species is not obviously homologous with the conserved complement. The family Macropodidae (kangaroos and wallabies) is cytologically the most diverse group of marsupials. This chromosome differentiation in a group of species with rather similar biological roles, and between many of which viable but sterile hybrids form, is a striking contrast with the conserved complement and diversity of biological adaptations found in the majority of marsupials.

A number of species from different genera of macropods have a morphologically similar chromosome complement to that of Thylogale billardieri, and this complement, derived from the conserved complement, has been postulated to be ancestral for this family of marsupials. Using the ten G-banded autosomes of this species as "building blocks" it has been possible to construct the chromosome complements of all sixteen species of macropods that have been studied (Rofe, 1979). The predominant pattern of chromosome evolution has been one of fusion.

This group of marsupial species with considerable change in chromosome number and with variations in chromosome morphology between the different complements, provides a sharp contrast to the conserved complement. Yet G-banding analysis shows that both groups are using very similar chromosomal elements. The constraints which were operating on the conserved complement have clearly been lost. However, there is some evidence that, not surprisingly, a new set of constraints are operating upon the chromosomal phenotypes of these species. Two special restrictions of the fusion pattern can be suggested as evidence for this.

The first of these is shown by the nine species with 2n=16. Their chromosome complements which upon general chromosome staining appear to be similar, reveal upon G-banding to consist of three different groups. Each group has three fusions and the six chromosomes involved are the same. However, the chromosome combinations involved in the fusions are different in each group. On the argument that selective processes are involved in arriving at these complements, it appears that selection has favoured morphological similarity and a reduction in chromosome number rather than specific fusions as such. This is an example of what White (1973, 1975) called "karyotypic orthoselection". It is, of course, similar to, but a much less stringent constraint than those postulated to be responsible for favouring the conserved complement.

The second pattern is shown by the size of the chromosomes produced by chromosome fusion. If the chromosomes which have combined and been incorporated into chromosome complements were to do so at random, then one might expect to have some chromosome complements with very large chromosomes. However, as chromosome number is reduced in this group and chromosome size generally is increased, there is no evidence of disproportionate 'growth' of single chromosomes. Rather the large chromosomes in all the complements generally have much the same size. This certainly suggests the operation of selection upon the complement as a whole, rather than chance effects. Since much the same G-banding elements are present in the chromosomes of the Macropodidae as are present in the chromosomes of the conserved complement, it is interesting that the 2n=14 species of macropods do not resemble the conserved complement either by G-band combination or by general morphology. It is clear that in this group and in others where the conserved complement is no longer found, that a new set of constraints are operating upon the karyotype and a different product or products is favoured.

As part of the broad overview of marsupial chromosome

evolution, we have made use of our knowledge of the G-banding homologies in marsupials in a study of the variation that exist in recombination at meiosis (Sharp and Hayman, unpublished). The chromosome number and chiasma frequency provide a measure of the number of segregating units at meiosis. Darlington (1958) suggested that these features were combined into a genetic system with components of the natural history of an organism such as, for example, the degree of inbreeding and the generation time. This idea has been explored largely in plants and patterns of association have been recognised.

Little data exists about chiasma frequency in mammals, and few studies of this sort have been reported. We have collected data on the chiasma frequency in males of thirty three species of marsupial. There are obvious shortcomings in this study since the number of animals studied from any species is not large. Further, only males were examined and there are substantial differences in recombination between the sexes in at least one species of marsupial (Bennett et al., 1986). However, the opportunity that the conserved complement offers for comparative studies of chiasmata frequency with other biological features justified the collection even of limited data. The results are shown in Tables 2 and 3.

There is a two-fold range in chiasma frequency over the conserved complement. Highly significant correlations occur between chiasma frequency and the metrics of size and rate of development. The correlation with the number of pouch young reared is negative and non-significant. However, it is likely that if the number of young reared by each female in each year was used as an index of reproductive capacity, there would be a significant correlation since some of the smaller dasyurids and the peramelids have multiple litters each year.

In these species the correlations suggest that the highest chiasma frequency and hence the greatest degree of genetic variability in the gametic output exists in those species with the largest size and slowest development time. It will be of interest to compare these results with comparable data from other groups of mammals.

TABLE 2 Chiasmata frequency in species with the conserved 2n=14 complement.

Species	No. of individuals	Av. chiasmata per cell
DASYURIDAE		
Sminthopsis crassicaudata	17	13.64 ± 1.6
Sminthopsis murina	1	13.50 ± 1.0
Dasyuroides byrnei	4	17.40 ± 1.8
Dasyurus hallucatus	1	18.75 ± 1.7
Dasyurus viverrinus	3	18.93 ± 1.7
Sarcophilus harrisii	3	20.83 ± 1.9
PERAMELIDAE		
Perameles gunnii	4	12.66 ± 1.3
Perameles nasuta	2	13.98 ± 1.9
Isoodon macrourus	1	16.27 ± 1.9
Isoodon obesulus	3	16.30 ± 1.5
VOMBATIDAE		
Lasiorhinus latifrons	2	22.43 ± 1.8
Vombatus ursinus	1	22.20 ± 2.5

At least 20 cells per individual were scored

TABLE 3 Spearman correlations between various features and chiasma frequency in species with the conserved 2n=14 complement.

Feature	N	R_s	Significance
Head-body length	12	+0.69	*
Weight	12	+0.75	**
Gestation time	10	+0.80	**
Birth → weaning time	12	+0.88	***
No. of pouch young	12	-0.20	NS
♀ age at 1st breeding	12	+0.88	***
DNA content	9	-0.13	NS

* 5% level, ** 1% level, *** 0.1% level; NS = not significant

REFERENCES

Air, G. M., Thompson, E. O. P., Richardson, B. J. and G. A. Sharman 1971.Amino-acid sequences of kangaroo myoglobin and haemoglobin and the date of marsupial-eutherian divergence. Nature 229, 391-394.

Bennett, J. H., Hayman, D. L. and Hope, R. M. 1986.Novel sex differences in linkage value and meiotic chromosome behaviour in a marsupial. Nature, in press.

Darlington, C. D. 1958.The Evolution of Genetic Systems. Oliver and Boyd: Edinburgh/London.

Hayman, D. L. 1977.Chromosome number - constancy and variation. In The Biology of Marsupials, eds. B. Stonehouse and D. Gilmore, 27-50. Macmillan Press: London.

Hayman, D. L., Kirsch, J. A. W., Martin, P. G. and P. F. Waller 1971.Chromosomal and serological studies of the Caenolestidae and their implications for marsupial evolution. Nature 231, 194-195.

Hayman, D. L. and P. G. Martin 1969.Cytogenetics of marsupials. In Comparative Mammalian Cytogenetics, ed. K. Benirschke, 191-217. Springer Verlag: New York.

Hayman, D. L. and P. G. Martin 1974.Monotremata and Marsupialia: Mammalia 1, Vol. 4: Chordata 4. In Animal Cytogenetics ed. B. John, pp. 110. Gebruder Borntraeger: Berlin-Stuttgart.

Hayman, D. L. and P. J. Sharp 1982.The chromosomes of Tarsipes spencerae Gray (Marsupialia). Aust. J. Zool. 30, 799-803.

Keast, A. 1977.Historical biogeography of the marsupials. In The Biology of Marsupials, ed. D. Hunsaker, 1-50. Academic Press.

Kirsch, J. A. W. 1977.The comparative serology of Marsupialia, and a classification of marsupials. Aust. J. Zool. Suppl. Series, 52, 1-152.

Kirsch, J. A. W. and J. H. Calaby 1977.The species of living marsupials: an annotated list. In The Biology of Marsupials, ed. B. Stonehouse and D. Gilmore, 9-26. Macmillan: London.

Macgregor, H. C. 1982.Big chromosomes and speciation amongst Amphibia. In Genome Evolution, ed. G. A. Dover and R. B. Flavell, 326-341. The Systematics Association Special Volume No. 20. Academic Press.

Matthey, R. 1973. The chromosome formulae of eutherian mammals. In Cytotaxonomy and Vertebrate Evolution, ed. A. B. Chiarelli and E. Capanna, 531-616. Academic Press: London/New York.

McClintock, B. 1941.Spontaneous alterations in chromosome size and form in Zea mays. Cold Spring Harbor Symposium in Quantitative Biology 9, 72-81.

Muller, H. J. 1940.Bearings of the Drosophila work on systematics. In The New Systematics, ed. J. Huxley. Oxford University Press.

Rofe, R. H. 1979.G-banding and Chromosomal Evolution in Marsupials. Ph.D. thesis, University of Adelaide, South Australia.

Rofe, R. and D. Hayman 1985.G-banding evidence for a conserved complement in the Marsupialia. Cytogenet. Cell Genet. 39, 40-50.

Sharman, G. B. 1973.Marsupial taxonomy and phylogeny. Australian Mammalogy 1, 137-154.

Sharman, G. B. 1974.The chromosomes of non-eutherian mammals. In Cytotaxonomy and Vertebrate Evolution, eds. A. B. Chiarelli and E. Capanna, 485-530. Academic Press: New York.

White, M. J. D. 1973.Animal Cytology and Evolution, 3rd edition. Cambridge University Press.

White, M. J. D. 1975.Chromosome repatterning - regularities and restrictions. Genetics 79, 63-72.

Woodburne, M. O. and W. J. Zinsmeirer 1982.Fossil land mammal from Antarctica. Science 218, 284-285.

Young, G. J., Marshall Graves, J. A., Barbieri, I., Woolley, P. A., Cooper, D. W. and M. Westerman 1982.The chromosomes of Dasyurids. In Carnivorous Marsupials, ed. M. Archer, 783-795. Royal Zoological Society of New South Wales, Sydney, Australia.

Comparative gene mapping and primate evolution

N. Créau-Goldberg, C. Cochet, C. Turleau and J. de Grouchy

INSERM U.173, CNRS U.A. 119, Hôpital Necker, Paris, France

Chromosome homologies defined by banding techniques have allowed the elucidation of the primate chromosomal phylogeny (Turleau et al. 1972, Grouchy et al. 1978, Dutrillaux 1979, Yunis and Prakash 1982). The correspondance of these homologies and the homology of gene content was shown by the first comparative mappings between man and the great apes obtained by somatic cell genetics (Finaz et al. 1973-1975-1977, Pearson et al. 1977). When more distantly related species are studied, homologies are more difficult to define. Nevertheless, Dutrillaux's group (Dutrillaux 1979, Dutrillaux et al. 1986 for review) has suggested most of the rearrangements having occurred during primate evolution. Only gene mapping in these species could confirm or infirm these homologies and demonstrate homologies not defined by chromosome banding, as for the gibbon.

Most gene assignments in primates were obtained by somatic cell genetics. As for the human gene map, the first genes localized code mostly for enzymes which are expressed in hybrid cells. The development of molecular hybridization techniques allows the investigation of genes not expressed in hybrid cells and a broader understanding of chromosome evolution by the analysis of the degree of gene homology, of gene restriction maps and of the presence of pseudogenes.

The COL1A1 gene affords an example of this type of investigation. It was localized in the baboon and the African green monkey (Cercopithecoïdae) using a cDNA probe covering approximately

75 % of the human gene (Chu et al. 1982). These localizations together with those of syntenic genes have demonstrated the conservation of a great part of chromosome HSA17 during primate evolution (Créau-Goldberg et al. 1984). The same probe was hybridized with DNAs of different primate species digested by six different restriction enzymes. The results showed a good conservation of restriction sites when compared to human DNA. No modification was observed between man, chimpanzee and gorilla for the enzymes studied. Conversely RFLPs have been observed in the Cercopithecoïdae (with respect to man) and seem to be species-specific (fig. 1). The construction of complete restriction maps should bring more information on the differences between these genes, as it was the case for the globin genes (Barrie and Jeffreys 1981, Martin et al. 1983).

I will discuss five instances showing the importance of gene mapping in solving chromosomes homologies. I will particularly insist on the following species with regard to their place in the phylogenetic tree : (1) the gibbon (Hylobatidae) which belongs to the Hominidae but shares no evident homology with the human karyotype. This characteristic is wide-spread in all species or subspecies of the family Hylobatidae, (2) the capuchin monkey (Cebidae) whose karyotype has been defined the closest to the ancestral karyotype of all Simians and (3) the mouse lemur whose karyotype is close to the ancestral karyotype of all primates.

1-HSA1 (fig. 2)

The main HSA1 markers studied were ENO1 (pter-p36.31), PGD (p36.2-p36.13), FUCA1 (p34), PGM1 (p22.1), PEPC (q25 or q42), GUK1 (q32-q42), FH (q42.1). These markers have been assigned in the Pongidae on homologs of HSA1 (Finaz et al. 1973-1975-1977, Pearson et al. 1978). The genes are still syntenic in the baboon and rhesus (Finaz et al. 1977, Lalley et al. 1979, Garver et al. 1978). There is a disruption of synteny in the A.g. monkey, confirming the fission between 1p and 1q (Van Cong N. et al. 1978, Garver et al. 1978). In fact, the HSA1p is conserved in the whole phylum since the mouse lemur (Cochet et al. 1982), in the capu-

Figure 1 :
Southern blot of Hind III digested DNAs hybridized with COL1A1 probe : Hf 404.

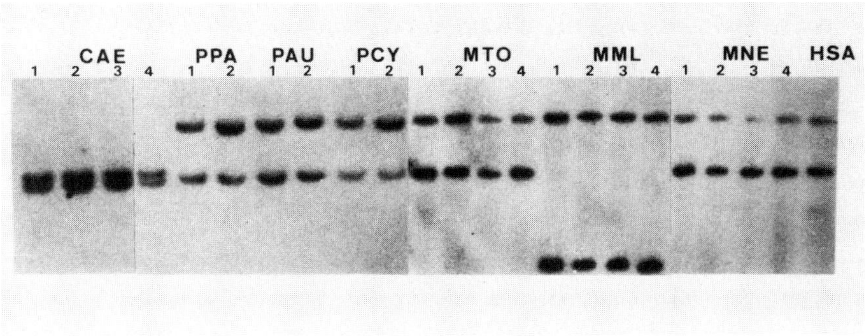

Legend : CAE: A.g. monkey, PPA: Guinea baboon, PAU: olive baboon, PCY: yellow baboon, MTO: Moor macaque, MML: rhesus monkey, MNE: pigtailed macaque, HSA: man.

Figure 2 :
Comparative mapping for HSA1.

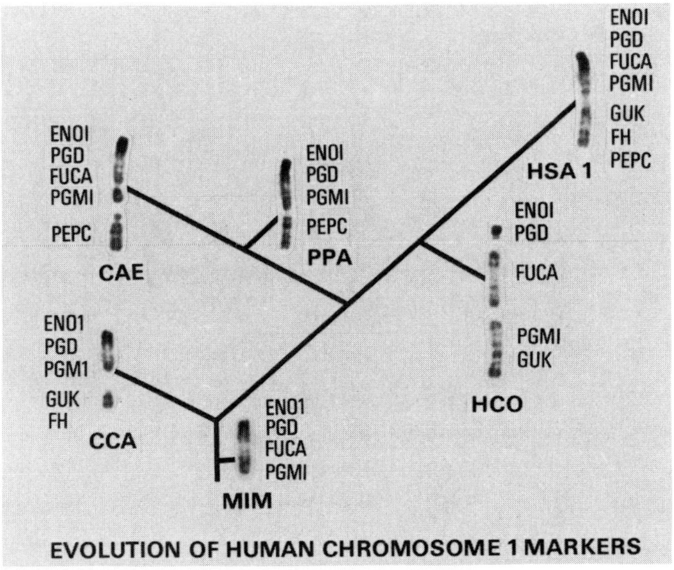

chin monkey (Créau-Goldberg et al. 1981) and also in the owl monkey (Cebidae) (Ma 1983). More complex homologies have been defined, only by gene mapping, in the gibbon (Turleau et al. 1983). There are at least 3 different chromosomes carrying HSA1p markers. Only one HSA1q marker was studied and is syntenic with PGM1. Further localizations are needed to define the rearrangements which have occurred. HSA1q markers in the most ancestral species show that at the origin of primates exist at least two or more partial homologs : in the mouse lemur, GUK1 and PEPC are asyntenic but not assigned. In the capuchin monkey, GUK1 and FH are carried by a small chromosome which can't be the entire homolog of HSA1q. Only FH is localized in the owl monkey on a metacentric chromosome.

A comparison with other mammalian species shows that PGD-ENO1-PGM1 synteny is found in many : mouse, rat hamster, cat, dog, sheep and mink (Comparative Mapping Committee, HGM8, 1985). Asynteny of these markers has been shown only in the cow (Heuertz and Hors-Cayla 1981) and the rabbit (Soulié and de Grouchy 1983). Little is known about HSA1q markers in these species. This confirms that the HSA1p synteny is present very early in evolution, but that HSA1q markers were disseminated between different chromosomes in the ancestral species.

2-HSA6 (fig. 3)

Five genes were mainly studied among HSA6 markers : HLA (p21.3), GLO1 (p21.31-p21.1), ME1 (q12), PGM3 (q12) and SOD2 (q21). HSA6 is highly conserved in the Pongidae and disrupted in the Hylobatidae. These markers are also on a single chromosome in the Cebidae ; and in the mouse lemur, 6p and 6q markers are on an acrocentric chromosome.

A confusing situation exists for the baboon (PPA) and rhesus (MML) maps, though the karyotypes of these two species have been defined as homologous (Finaz et al. 1976). The results of our group (Créau-Goldberg et al. 1982a -1983-1985-1986) and Peter Lalley's (Thiessen and Lalley 1986) demonstrate that HSA6 markers are on PPA4 ,and HSA7 and 21 markers are on PPA3. Gene mapping in the rhesus by Estop et al. (1983) has brought HSA6 and 7 mar-

Figure 3 :
Comparative mapping for HSA6

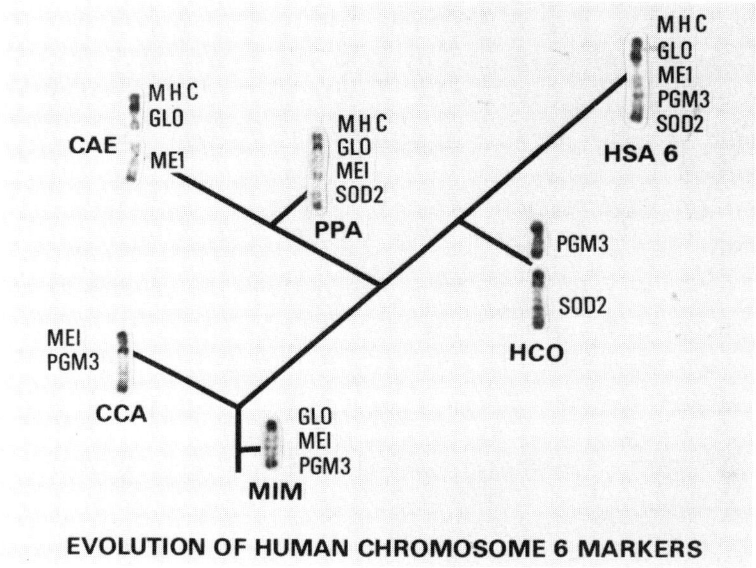

Figure 4 :
Comparative mapping for HSA12

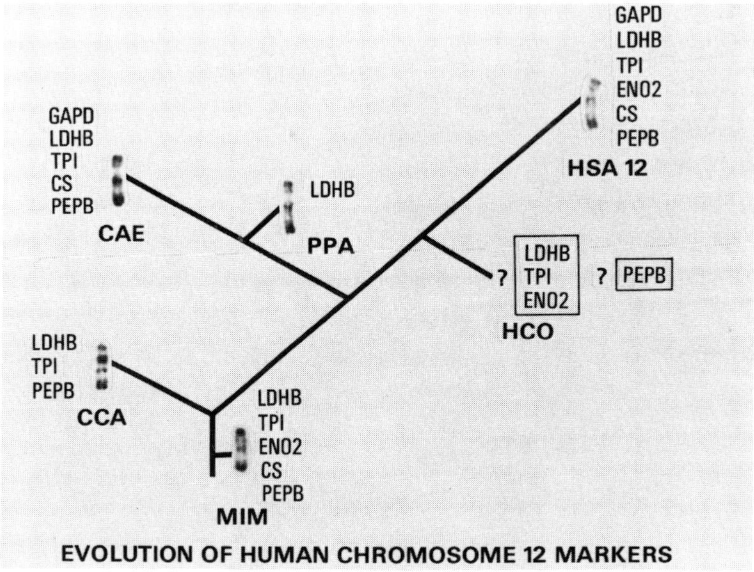

kers on MML2 which has been considered homologous to PPA3. This discrepancy may require gene mapping in other individuals of the same species to confirm or infirm these results. The baboon map is also in complete agreement with the localizations we obtained in the A.g. monkey. MHC (for HLA) and GLO1 on one part and ME1 on the other part give another demonstration of the evolution by fission with regard to baboon chromosomes (Créau-Goldberg et al. 1985-1986).

The synteny of HSA6 markers has been also shown in some other mammalian species, for instance the cat (O'Brien and Nash 1982) and the dog (Meera Khan et al. 1978-1984). In the mouse (Roderick and Davisson 1984), MHC and GLO1 are syntenic with SOD2 but asyntenic with ME1 and PGM3. ME1-PGM3 synteny is the most conserved in mammals.

3-HSA12 (fig. 4)

The main HSA12 markers studied are GAPD (p13), LDHB (p12.2-p12.1) TPI1 (p13), ENO2 and CS (p11-qter), and PEPB (q21). Homologs of this HSA chromosome are present in all primate species, except in the gibbon because it has undergone few rearrangements (Turleau et al. 1983). Markers of 12p and 12q are still syntenic in the A.g. monkey (Garver et al. 1978) and in the capuchin monkey (Créau-Goldberg et al. 1981). In the mouse lemur, they are on an acrocentric chromosome. For this species, the results obtained in hybrids reveal a close linkage of ENO2 and TPI1, LDHB and CS, demonstrating that the gene order is different from that of HSA12 (Cochet et al. 1982). This synteny across the centromere is also conserved in the cow (Heuertz and Hors-Cayla 1981). Only 12p synteny is found in the mouse (Roderick and Davisson 1984) as in the gibbon, but not assigned in that species.

4-HSA14 and 15 (fig. 5)

The markers studied for HSA 14 are : NP (q13.1) and CKBB (q32) and for HSA15 : SORD (pter-q21), HEXA (q22-q25.1) and MPI-PKM2-IDH2 (q22-qter). Studies of these markers have shown their common evolution in primates, demonstrating that only a large part of

Figure 5 :
Comparative mapping for HSA14 and 15

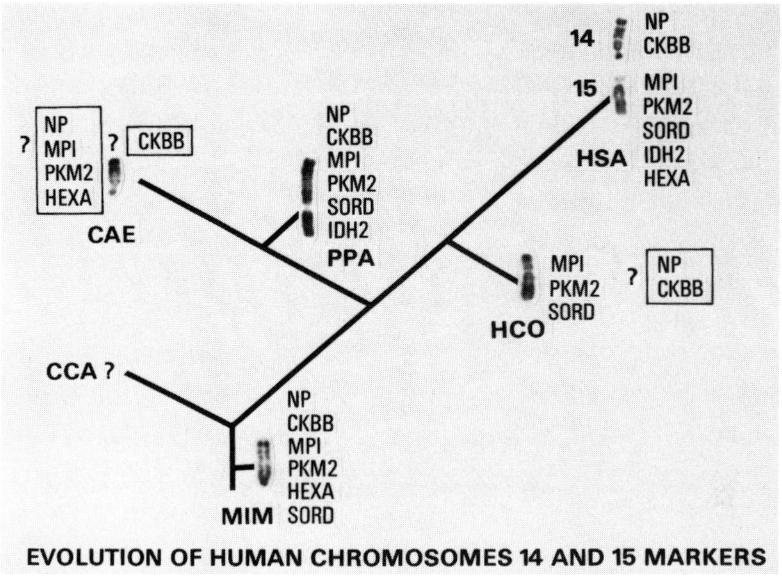

Table 1 :
Comparative mapping of DNA sequences

Probe	Man	Pongidae			Cercopithecoïdae	Lemurs	Authors
		PTR	GGO	PPY	(CAE - MML)		
DXS31	Xp223-pter	X	nt	nt	X	autosomal or pseudoaut.	KOENIG et al. 1984
	Yq11-qter	Y			-		
DXYS1	Xq	X	X	X	nt	nt	PAGE et al. 1984
	Yp	-	-	-			
DXYS8	Xq12-q22	X	nt	nt	X	nt	WEISSENBACH et al. 1984
	Y	-			-		
DXYS12	Xq13-q22	X	nt	nt	X	nt	KOENIG et al. 1985
	Y	?			-		
ASSP	X(2)	X(2)	X(1?)	X	PPA : X(1)	nt	DAIGER et al. 1985
	Y	Y	Y	-	-		
ACTL2	Yp11	Y	nt	nt	autosomal	autosomal	KOENIG et al. 1985

Legend : PTR: chimpanzee, GGO: gorilla, PPY: orangutan, CAE: A. g. monkey, MML: macaque, PPA: baboon, nt: not tested.

14q is ancestral.

In the mouse lemur, we found a chromosome similar to HSA14 but with NP, CKBB and also MPI, PKM2, HEXA and SORD. The localizations of these markers in man suggest that about half or more of HSA15 is translocated on HSA14. The same synteny was found in the owl monkey - it is not clear in the capuchin monkey - in the marmoset (Hink and Lalley 1985), but in this case only NP was studied. In the baboons and macaques, the homolog corresponds to HSA14+15. There is a fission in the A.g. monkey disrupting the synteny of NP and CKBB (Créau-Goldberg et al. 1982b). Because of its localization near the centromere of HSA14, NP is a good marker of the instability of this region. As it is suggested by the mapping in the A.g. monkey, the homolog of HSA14 does not carry this gene and has probably created a new centromere. In the superfamily Hominidae (chimpanzee, gorilla, orangutan, gibbon), HSA14 and 15 markers are asyntenic.

The synteny of NP with HSA15 markers has been also shown in the cat and the pig. The synteny MPI-PKM2 being conserved in most mammals except cow and sheep (Comparative Mapping Committee, HGM8, 1985).

5-HSAX

As in most species, the X is widely conserved in primates (Comparative Mapping Committee, HGM8, 1985 and Weber et al. this conference). But recent investigations with DNA sequences have revealed that they may be exchanged between chromosomes during evolution. In table 1, some of these sequences are presented. In the case of the three probes DXYS1, DXYS8 and DXYS12 - which detect X-Y almost identical sequences (99 % homology) outside the meiotic pairing region in man - an X towards Y exchange seems to have occurred only in the human phylum. The X-Y exchange for DXS31 seems to be more ancient, maybe before the Pongidae, which is in agreement with the 80 % homology between X and Y sequences (Koenig et al. 1984). The two other sequences represent instances of gene families widely dispersed in the genome. Argininosuccinate synthase (ASS) whose functional gene is localized on HSA9 has many pseudogenes (ASSP) assigned on several chromosomes.

Two of them are on the X and one on the Y. Daiger et al. (1983-1986) have shown that the Y pseudogene is present in the chimpanzee and gorilla while only the chimpanzee has two pseudogenes on the X, as man. For ACTL2 which is a flanking sequence of the Y actin gene, the sequence is present only in the chimpanzee on the Y and autosomal in the Cercopithecoïdae and Lemurs.

These few examples demonstrate exchanges between X and Y and autosomes at different stages in primate evolution. As more genes will be studied in primates, insertions of such sequence in their genomes, that chromosome banding cannot detect, should be shown, as the repetitive elements.

Comparative gene mapping in primates has another important aspect, since many models of human diseases are developed in primates (i.e. association of dyslipoproteinemia and atherosclerosis in baboons, immunogenetic aspects of mother-fetus relationships in macaques, infection by HB virus in chimpanzee). As for human diseases, linkage analysis with RFLPs can be determined. This sort of analysis in models where environmental and genetic variables can be manipulated, may be a powerful approach to identify the respective role of these variables and their interaction in the etiology of diseases.

REFERENCES

Barrie, P. A. and Jeffreys, A. J. 1981. Evolution of the β-globin gene cluster in man and the primates. J. Mol. Biol. 149, 319-336.

Chu, M. L., Myers, J. C., Bernard, M.P., Ding, J.F. and Ramirez, F. 1982. Cloning and characterization of five overlapping cDNAS specific for the human pro α1(I) collagen chain. Nucl. Acid Res. 10, 5925-5934.

Cochet, C., Créau-Goldberg, N., Turleau, C., Grouchy, J. de 1982. Gene mapping of *Microcebus murinus* (Lemuridae). A comparison with man and *Cebus capucinus* (Cebidae). Cytogenet. Cell Genet. 33, 213-221.

Comparative Mapping Committee (Lalley, P.A. and Mc Kusick, V.A.) 1985 HGM8, Helsinki Conference (1985). Cytogenet. Cell Genet. 40, 536-566.

Créau-Goldberg, N., Cochet, C., Turleau, C., Grouchy, J. de 1981. Compared gene mapping of man and *Cebus capucinus*. A study of 23 enzymatic markers. Cytogenet. Cell Genet. 31, 228-239.

Créau-Goldberg, N., Turleau, C., Cochet, C., Grouchy, J. de 1982a. Comparative gene mapping of the baboon (*Papio papio*) and man.

Ann. Génét. 25, 14-18.
Créau-Goldberg, N., Turleau, C., Cochet, C. and Grouchy, J. de 1982b. Carte génique des primates. Homologies chromosomiques avec le caryotype humain. First European Congress on Cell Biology (Paris, July 1982).
Créau-Goldberg, N., Turleau, C., Cochet, C. and Grouchy, J. de 1983. New gene assignments in the baboon and new chromosome homoeologies with man. Ann. Génét. 26, 75-78.
Créau-Goldberg, N., Turleau, C., Cochet, C., Huerre, C., Junien, C. and Grouchy, J. de 1984. Conservation of the human COL 1A1-TK-GAA synteny and homoeologous assignment in the African green monkey and the baboon (Cercopithecoïdae). Human. Genet. 68, 333-336.
Créau-Goldberg, N., Cochet, C., Turleau, C., and Huerre-Jeanpierre, C., Junien, C. and Grouchy, J. de 1985. Primate gene mapping : the genes of collagen I and some other genes. HGM8, Helsinki Conference (1985). Cytogenet. Cell Genet. 40, 610.
Créau-Goldberg, N. and Cochet, C. Unpublished results.
Daiger, S.P. and Hoffman, N.S. 1983. Comparison of Y chromosome DNA sequences between humans and chimpanzees. Genetics 104, 20.
Daiger, S.P., Goode, M.E. and Trowbridge, B.D. 1986. Evolution of nuclear families. Abs. Genetic Research with non human primates : serving the need of mankind, San Antonio, Texas.
Dutrillaux, B. 1979. Chromosomal evolution in primates : tentative phylogeny from Microcebus murinus (Prosimian) to man. Hum. Genet. 48, 251-314.
Dutrillaux, B. et al. Evolution chromosomique chez les primates, les carnivores et les rongeurs. 1986, Mammalia 50, numéro spécial.
Estop, A.M. Garver, J.J., Egozcue, J., Meera Khan, P. and Pearson, P.L. 1983. Complex chromosome homologies between the Rhesus monkey (Macaca mulatta) and man. Cytogenet. Cell Genet. 35, 46-50.
Finaz, C., Turleau, C., Grouchy, J. de, Nguyen, V.C., Rebourcet, R., Frézal, J. and Chavin-Colin, F. 1973. Comparison of man and chimpanzee syntenic groups by cell hybridization. Preliminary report. Biomedicine 19, 526-531.
Finaz, C., Cochet, C., Grouchy, J. de, Nguyen, V.C., Rebourcet, R. and Frézal, J. 1975. Localisation génique chez le chimpanzé (Pan troglodytes) : comparaison avec la carte factorielle de l'homme (Homo sapiens). Ann. Génét. 18, 169-177.
Finaz, C., Nguyen, V.C., Cochet, C., Frézal, J. and Grouchy, J. de 1977. Histoire naturelle du chromosome 1 chez les primates. Ann. Génét. 20, 85-92.
Garver, J., Pearson, P.L., Estop, A., Dijksman, T., Wijnen, L., Westerveld, A. and Meera Khan, P. 1978. Gene assignments to the presumptive homologs of human chromosomes 1, 6, 11, 12 and X in the Pongidae and Cercopithecoidae. HGM4, Winnipeg Conference (1977). Cytogenet. Cell Genet. 22, 564-569.
Grouchy, J. de, Turleau, C. and Finaz, C. 1978. Chromosomal phylogeny of the primates. Ann. Rev. Genet. 12, 289-328.
Heuertz, S. and Hors-Cayla, M.C. 1981. Cattle gene mapping by somatic cell hybridization ; study of 17 enzyme markers. Cytogenet. Cell Genet. 30, 137-145.
Hink, L.A. and Lalley, P.A. 1985. A biochemical genetic map of

the cotton-topped marmoset, Saguinus oedipus : a comparison with man, mouse and other primates. HGM8, Helsinki Conference (1985). Cytogenet. Cell Genet. 40, 651.

Koenig, M., Camerino, G., Heilig, R. and Mandel, J.L. 1984. A DNA fragment from the human X chromosome short arm which detects a partially homologous sequence on the Y chromosomes long arm. Nucl. Acid. Res. 12, 4097-4109.

Koenig, M., Moisan, J.P., Heilig, R. and Mandel, J.L. 1985. Homologies between X and Y chromosomes detected by DNA probes : localisation and evolution. Nucl. Acid. Res. 13, 5485-5501.

Lalley, P.A., Dixon, D. and Held, M.B. 1979. Gene mapping studies in the baboon (Papio papio) using somatic cell hybrids. HGM5, Edinburgh Conference (1979). Cytogenet. Cell Genet. 25, 178.

Ma, N.S.F. 1983. Comparative gene mapping of the owl monkey, Aotus, with karyotypes V (2n = 46) and VI (2n = 49,50). Cytogenet. Cell Genet. 35, 117-123.

Martin, S.L. Vincent, K.A. and Wilson, A.C. 1983. Rise and fall of the delta globin gene. J. Mol. Biol. 164, 513-528.

Meera Khan, P., Brahe, C. and Wijnen, L.M.M. 1984. Gene map of dog : six conserved and three disrupted syntenies. HGM7, Oslo Conference (1983). Cytogenet Cell Genet 37, 537-538.

Meera Khan, P., Vriesendorp, H.M., Saison, R., Volkers, W.S., Los, W.R.T. and Doppert, B.A. 1978. Linkage between PGM3 and the genes determining the major histocompatibility complex (MHC) in Canis familiaris (the domestic dog). Cytogenet. Cell Genet. 22, 585-587.

Myers, J.C., Chu, M.L., Faro, S.H., Clark, W.J., Prockop, D.J. and Ramirez, F. 1981. Cloning a cDNA for the pro α 2 chain of human type I collagen. Proc. Natl. Acad. Sci. USA 78, 3516-3520.

Nguyen, V.C., Weil, D., Finaz, C., Rebourcet, R., Cochet, C., Grouchy, J. de and Frézal, J. 1978. Localisation du gène α-FUC sur 1p chez l'homme et le chimpanzé et sur le 4 chez le cercopithèque. Ann. Génét. 21, 41-46.

O'Brien, S.J. and Nash, W.G. 1982. Genetic mapping in mammals: chromosome map of domestic cat. Science 216, 257-265.

Page, D.C., Harper, M.E., Love, J. and Botstein, D. 1984. Occurrence of a transposition from the X-chromosome long arm to the Y chromosome short arm during human evolution. Nature 311, 119-123.

Pearson, P.L., Garver, J.J., Estop, A., Dijksman, T.M., Wijnen, L.M.M. and Meera Khan, P. 1978. Gene assignments to the presumptive homologs of human chromosome 2, 9, 13, 14 and 15 in the Pongidae and Cercopithecidae. Human Gene Mapping 4, Winnipeg Conference (1977). Cytogenet. Cell Genet. 22, 588-593.

Roderick, T.H. and Davisson, M.T. 1984. The linkage map of the mouse Mus musculus. Genetic Maps 3, 343-355.

Soulié, J. and Grouchy, J. de 1983. New gene assignments in the rabbit (Oryctolagus cuniculus). Comparison with other species. Human Genet 63, 48-52.

Thiessen, K.M. and Lalley, P.A. 1986. New gene assignments and syntenic groups in the baboon (Papio papio). Cytogenet. Cell Genet. 42, 19-23.

Turleau, C., Grouchy, J. de and Klein, M. 1972. Phylogénie chromosomique de l'homme et des primates hominiens (Pan troglodytes, Gorilla gorilla, et Pongo pygmaeus). Essai de reconstitution du caryotype de l'ancêtre commun. Ann. Genet. 15, 225-240.

Turleau, C, Créau-Goldberg, N., Cochet, C. and Grouchy, J. de 1983. Gene mapping of the gibbon. Its position in primate evolution. Hum. Genet. 64, 65-72.

Weissenbach, J., Geldwerth, D., Guellaen, G., Fellous, M. and Bishop, C. 1984. Sequences homologous to the human Y chromosome detected in the female genome. HGM7, Oslo Conference (1983). Cytogenet. Cell Genet. 37, 604.

Yunis, J.J. and Prakash, O. 1982. The origin of man : a chromosomal pictorial legacy. Science 215, 1525-1530.

Laboratoire de Cytogénétique Humaine et Comparée, INSERM U.173, CNRS EA.119, Hôpital Necker-Enfants Malades, 149 Rue de Sèvres, 75743 Paris Cedex 15, France.

CHROMOSOMES AND MALIGNANT CHANGES

Chromosomal abnormalities in malignant lymphoid cell proliferations: recent advances

R. Berger

INSERM U.301, Hôpital Saint-Louis, Paris 75010, France

The coincidence of chromosomal breakpoints of acquired chromosomal rearrangements of cancer cells with protooncogene localizations has brought new insights on the pathogenetic mechanisms of cancerogenesis.Although the Burkitt lymphoma (BL) has been one of the first examples of human malignancy studied with molecular biology and molecular cytogenetic techniques,we now know other examples of malignant lymphoid cell proliferations illustrating this new approach of cancer study.The recognition of new chromosomal abnormalities and of new genes or DNA sequences has brought significant advances in this field since the last few years.

Cytogenetics of malignant lymphoid cell proliferations

Various nonrandom chromosomal changes have been reported in malignant lymphoid cell proliferations (Sandberg 1980,Berger et al.1985,Le Beau and Rowley 1986,Pedersen-Bjergaard et al.1986). As in other malignant cells,they may be numerical,structural,or both numerical and structural.Chromosomal abnormalities are clonal but generally differ in acute lymphoblastic leukemia (ALL) and in malignant lymphoma (ML) as far as complexity is concerned. The karyotypes of malignant lymphoma cells are often complex,with several changes and sometimes variation from cell to cell.However these cells share common abnormalities defining clonality. Since several years it has been found that specific chromosome changes were related to more or less specific cytologic or histological

Table 1.Chromosomal rearrangements in malignant lymphoid cell proliferations.

Chromosomal abnormalities	Diseases	Gene localization	
t(1;14)(p32;q11)	T-cell ALL	L-myc	TcRα
t(1;19)(q23;p13)	Pre B ALL		insulin receptor
t(2;14)(p11p13;q32)	ML,CLL		IgH
t(4;11)(q21;q23)	ALL or monocyte precursor L.		c-ets 1
del(6)(p23p24) or t	T-cell ML		
del(6)(q21-qter)	ALL,ML,ATL		
t(8;10)(q22;p14)	Lymphoproliferative disorders		
t(8;14)(q24;q11)	T-cell ALL,CLL	c-myc	TcRα
t(8;14)(q24;q32)	BL,ALL L3,ML	c-myc	IgH
t(8;22)(q24;q11)	BL,ALL L3	c-myc	Igκ
t(2;8)(p11p12;q24)	BL,ALL L3	Igλ	c-myc
del(9)(p21p22)	pre B ALL	interferon	
t(10;14)(q24;q11)	T-cell ALL		TcRα
inv(14)(q11q32)	T-cell CLL,ML	TcR α	IgH
t(14;14)(q32;q11)	T-cell ALL or ML in ataxia-telan--giectasia	TcR α	IgH
t(14;18)(q32;q21)	Pre B or B ML	IgH	bcl 2
t(11;14)(p13;q11)	T-cell ALL,CLL		TcRα
t(11;14)(q13;q32)	pre B or B ALL,ML	bcl 1	IgH
t(11;14)(q23;q32)	pre B or B ALL,ML	c-ets 1	IgH

ALL:acute lymphoblastic leukemia.ALL,L3:acute lymphoblastic leukemia,Burkitt type.ATL: adult T-cell leukemia-mymphoma.BL: Burkitt lymphoma.CLL:chronic lymphocytic leukemia.ML:malignant lymphoma.IgH,Ig ,Ig :Immunoglobulin heavy chain, light chain, light chain.TcR :T-cell receptor alpha chain.

types of proliferation (table 1).However these correlations between cytology and cytogenetics may be questioned in some cases because of the diversity of malignant lymphoid cell proliferations and of difficulties in classifications.An example is the t(8;14)(q24;q32) translocation that is generally observed in Burkitt lymphoma-leukemia but which may also be present in other types of ML (Sigaux et al.1984,Yunis et al 1984,Levine et al.1985). Similarly the t(14;18)(q32;q21) translocation is associated with follicular lymphoma but it has also been found in some diffuse ML.The common point is that they were formed by cells of follicu--lar center cell origin.Chromosomal breakpoints at band 14q32 where the immunoglobulin heavy chain gene cluster (IgH) is located are particularly frequent in B-cell malignant proliferations,ALL, ML,multiple myeloma (Dewald et al.1985),chronic lymphocytic leukemia (CLL)(Gahrton et al.1985),and hairy cell leukemia (Berger et al.1985,Brito-Babapulle et al.1986).The fact that this band was also involved in rearrangements of T-cell proliferations (Nowell et al.1986) must be underlined.

A trisomy 12 has been reported to be nonrandomly associated with B-cell CLL (Gahrton et al.1985) and in ML,small lymphocytic. Similarly a trisomy 7 was associated with adult T-cell leukemia lymphoma in Japan (Ueshima et al.1981).The trisomies are also frequent in other ML,mainly trisomy 12,and their biological significance is poorly understood.In a survey of 258 ML cases trisomy 12 was found to be distributed among all groups of the classification, whereas isolated trisomy 12 was associated with ML,small lymphocytic only (Berger et al.1985).It suggests that trisomy 12 has a special role in this form of ML.

Finally a distinction has been made between "primary" chromoso--mal changes,i.e. close to a causative event,and "secondary" changes which appeared in the course of evolution of the disease. The majority of cases with the same type of proliferation have the first ones in common (Ph1 in chronic myelocytic leukemia, t(15;17) in acute promyelocytic leukemia,t(8;14) and variant in BL).The secondary changes,although often nonrandom, are superimpo--sed to the others and vary more largely in their incidence as well as in their specificity.For instance,a partial duplication of 1q ,often present in BL,has also been reported in other ML.

Cytogenetics,genes,and DNA.

Many chromosomal breakpoints involved in chromosomal abnormali--ties of cancer cells correspond to the localization of identified genes such as protooncogenes (table 1).

BL was the first human malignancy studied with both cytogenetic and molecular biology techniques. The finding of a correlation between the type of variant translocation and the type of immunoglobulin light chain expressed (Lenoir et al.1981),as well as the fact that the band 8q24 was a band rearranged in all the translocations described in BL-ALL,L3 (Berger and Bernheim 1985) have been an important step.The chromosomal localization of the protooncogene c-myc on 8q24,of IgH gene cluster on 14q32,and of immunoglobulin kappa and lambda light chain genes respectively on 2p11p12 and 22q11,i.e. chromosomal breakpoints involved in the translocations led to molecular investigation of Burkitt types of proliferations.Summarizing an important amount of work,it can be said that in t(8,14) translocation,c-myc is translocated onto the rearranged chromosome 14 and that the IgH cluster is split in two parts.One part (constant region) remains on the 14,and the other (variable region) is transposed onto the rearranged 8.Precise nucleotidic breakpoints vary from case to case ,but are mainly located in physiologically rearranging regions of immunoglobulin genes (J,S regions).The breakpoints in c-myc are more frequently located in the first exon,first intron or 5' flanking region in sporadic BL,and in the 5' side outside c-myc in endemic BL (Pelicci et al.1986).The variant translocations,t(2;8) and t(8;22), differ in that c-myc remains on the abnormal chromosome 8 and a part of DNA sequences coding for one of the immunoglobulin light chain (constant region of kappa and lambda chain respectively) is translocated onto this chromosome. The nucleotidic breakpoints vary on both chromosomes .On the chromosome 8, they are located ouside c-myc on its 3' side,in certain cases far from the gene (untill about 100 kb).It thus appears that the result of chromoso--mal rearrangement of BL is to place the protooncogene c-myc and immunoglobulin genes in contuguity resulting in a deregulation of the c-myc expression whatever the type of translocation (Leder et al.1983,Croce et al.1984). The molecular mechanisms of this deregulations are probably of several sorts such as increase of c-myc mRNA stability (Piechaczyk et al.1985) or making use of a new cryptic promotor (Murphy et al.1986).

The c-myc gene may also be involved in T-cell proliferations. A t(8;14)(q24;q11) translocation has been found in a variant subline derived by mutagenization from KE37, a cell line establi--shed from a T-cell ALL (Caubet et al.1985). The chromosomal breakpoints correspond to the bands where c-myc and T-cell receptor alpha chain gene (TcRα) are located.Molecular biology studies allowed our group to show that c-myc was abnormally expressed in this variant cell line,that the breakpoint on chromosome 8 was on the 3'side,close to the exon 3 of c-myc, and that TcRα gene was rearranged (Mathieu-Mahul et al.1985,1986). More recently,we had the opportunity to study a T-cell ALL in relapse with a t(8;14) identical translocation.Similarly c-myc and TcRα genes were rearranged.In this case however,neither the translocation nor the molecular rearrangements were observed at diagnosis.This suggests that these accidents were secondary events which occured in the course of the disease (Mathieu-Mahul et al.,unpublished).Other similar cases with the same molecular consequences have recently been reported (Erikson et al.1986, Shima et al.1986,McKeithan et al.1986),suggesting that this translocation is nonrandomly associated with T-cell malignancies. These new data may be compared to the variant translocations of BL because the localization of breakpoint was on the 3' side of c-myc exon 3 as well as the structure of TcRα gene ressemble that of immunoglobulins with physiological rearrangements during lymphoid cell maturation.

Recent studies have been made using the TcRα gene.Two of them concern a t(11;14)(p13;q32) translocation described in T-cell ALL (Williams et al.1984).In these cases, it has been shown that TcRα gene was split by the translocation,the Vα gene remained on the 14q- chromosome whereas Cα was translocated onto the 11p+. The localization on 11p13 is remarkable because it is the band involved in nephroblastoma with aniridia (Erikson et al.1985, Lewis et al.1985).

An inversion of chromosome 14,inv(14)(q11q32) has been described in T-cell CLL (Zech et al.1984) and later in T-cell various malignant proliferations (Zech et al.1986).It resulted in a transposition of the constant part of immunoglobulin mu heavy chain (C_μ) to a position close to the variable region (Vα) of TcRα gene on 14q11 and of Cα close to V_H region on 14q32. The rearranged DNA sequences $V_H J_\alpha C_\alpha$,on 14q32, was productive of a new hybrid immunoglobulin-T cell receptor polypeptide (Baer et al. 1985,Denny et al.1986).

All these examples have in common a rearrangement of a segment of one gene of the immunoglobulin gene superfamily.Two other translocations involving immunoglobulin heavy chain genes are also important for other reasons.The t(11;14)(q13;q32) translocation is associated with B-cell leukemia or lymphoma. The DNA segment corresponding to the breakpoint region on the chromosome 11 has been called bcl 1 (Erikson et al.1984).The other,the

the t(14;18)(q14;q21) translocation, associated with malignant lymphoma of follicular center origin, results in juxtaposition of constant region of IgH and of a segment of chromosome 18 called bcl 2 (Tsujimoto et al.1984).The bcl 2 sequence was expressed as mRNA in cancer cells (Tsujimoto et al.1985). These two translocations also have a special interest since their study led Tsujimoto et al. (1985,1985) to an hypothesis which might explain the nonrandomness of certain chromosomal translocations. In immunoglobulin genes, recombination signal sequences are normally located downstream the V region and upstream the J region.Each signal has an heptamer nucleotide sequence and a nonamer nucleotide sequence separated by spacers of 11 and 22 nucleotides. Because of homologies between heptamer and nonamer sequences, double structures can be formed allowing the recombination event to take place.A similar mechanism could explain the illegitimate recombination occuring in the formation of the translocations t(14;18)(q32;q21) and t(11;14)(q13;q32) as the presence of heptamer-nonamer sequences was shown on chromosomes 18 and 11 close to the chromosomal breakpoints. This model could be more general since DNA sequence homologies are known in genes coding for apparently unrelated proteins.

Recent work on chromosome translocations or inversions has clearly shown that they result in DNA sequence rearrangements. Some of the rearranged sequences are known as oncogenes,others as growth factor or receptor coding genes.It is not surprising that DNA sequences of the immunoglobulin gene superfamily,which are physiologically rearranged during B and T lymphocyte matura--tion, are also involved in pathological rearrangements related to chromosome abnormalities in malignant lymphoid cell proliferations.Other genes ,of course, may be involved as for instance the protooncogene c-ets 1 in the t(4;11) translocation associated to an ALL or a leukemia with monocyte precursor cells. It has been shown that c-ets 1 is translocated from the 11 to the 4 in this rearrangement (Sacchi et al.1986),but we do not know if the gene is deregulated or rearranged in this leukemia.

We now have to search to identify the other DNA sequences close to the chromosomal breakpoints in chromosomal abnormalities related to malignant cells in order to understand their molecular consequences as Fell et al.(1986) recently did for the t(2;14)(p13;q32) translocation associated to CLL.

References.

Baer R.,Chen K.-C.,Smith S.D. and Rabbitts T.H. 1985. Fusion of an immunoglobulin variable gene and a T cell receptor constant gene in the chromosome 14 inversion associated with T cell tumors. Cell 43,705-713.

Berger R. and Bernheim A. 1985. Cytogenetics of Burkitt's lymphoma -leukaemia : A review. In: Burkitt's lymphoma : A human cancer model. Lyon : IARC Scientific publications No.60.

Berger R.,Bernheim A. and Flandrin G. 1985. Lymphomes malins non Hodgkiniens et chromosomes.Actualités Hématologiques dix-neuvième série,48-56.Paris : Masson.

Berger R.,Bernheim A.,Valensi F. and Flandrin G. 1985. 14q- in two hairy cell leukemia patients. Cancer Genet.Cytogenet.16,91-93.

Berger R.,Bloomfield C.D. and Sutherland G.R. 1985. Report on the committee on chromosome rearrangements in neoplasia and on fragile sites.Human Gene Mapping 8. Cytogenet.Cell Genet.40, 490-535.

Brito-Babapulle V.,Pittman S.,Melo J.V.,Parreira L. and Catovsky D. 1986. The 14q+ marker in hairy cell leukaemia.A cytogenetic study of 15 cases.Leukemia Res.10,131-138.

Caubet J.-F.,Mathieu-Mahul D.,Bernheim A.,Larsen C.-J. and Berger R. 1985. Remaniement du locus du proto-oncogène c-myc dans une lignée cellulaire d'origine lymphoblastique T. C.R.Acad.Sc. Paris. 300,171-176.

Croce C.M.,Erikson J.,Nishikura K.,ar-Rushdi A.,Giallongo A., Rovera G.,Finan J. and Nowell P.C. 1984.235-242 in: Cancer Cells 2.Oncogenes and viral genes.Vande Woude G.F.,Levine A.J. Topp W.C. and Watson J.D. eds.Cold Spring Harbor : Cold Spring Harbor laboratory.

Denny C.T.,Yoshikai Y.,Mak T.W.,Smith S.D.,Hollis G.F. and Kirsch I.R. 1986. A chromosome 14 inversion in a T-cell lymphoma is caused by site-specific recombination between immunoglobulin and T-cell receptor loci. Nature 320,549-551.

Dewald G.W.,Kyle R.A.,Hicks G.A. and Greipp P.R. 1985. The clinical significance of cytogenetic studies in 100 patients with multiple myeloma,plasma cell leukemia,or amyloidosis. Blood 66,380-390.

Erikson J.,Finan J.,Tsujimoto Y.,Nowell P.C. and Croce C.M. 1984. The chromosome 14 breakpoint in neoplastic B cells with the t(11;14) translocation involves the immunoglobulin heavy chain locus. Proc.Natl.Acad.Sci.USA 81,4144-4148.

Erikson J.,Williams D.L.,Finan J.,Nowell P.C. and Croce C.M. 1985. Locus of the α-chain of the T-cell receptor is split by chromosome translocation in T-cell leukemias. Science 229,784-786.

Erikson J.,Finger L.,Sun L.,ar-Rushdi A.,Nishikura K.,Minowada J., Finan J.,Emanuel B.S.,Nowell P.C. and Croce C.M. 1986. Deregulation of c-myc by translocation of the α-locus of the T-cell receptor in T-cell leukemias. Science 232,884-886.

Fell H.P.,Smith G. and Tucker P.W. 1986. Molecular analysis of the t(2;14) translocation of childhood chronic lymphocytic leukemia. Science 232,491-494.

Gahrton G.,Juliusson G.,Robèrt K.-H.,Friberg K. and Zech L. 1985. Pathogenetic and clinical significance of chromosomal aberrations in B-cell chronic lymphocytic leukemia.45-53 in: Molecular Biology of Tumor Cells. Währen B.,Holm G.,Hammaström S. and Perlmann P. eds. New York : Raven Press.

Le Beau M.M. and Rowley J.D. 1986. Chromosomal abnormalities in leukemia and lymphoma : Clinical and biological significance. Adv.Hum.Genet.15,1-54. New York : Plenum.

Leder P.,Battey J.,Lenoir G.,Moulding C.,Murphy W.,Potter H., Stewart T. and Taub R. 1983. Translocations among antibody genes in human cancer.Science 222,765-7771.

Lenoir G.,Preud'homme J.L.,Bernheim A. and Berger R. 1981. Corrélation antre translocation variante et expression des chaînes légères d'immunoglobuline dans les lymphomes et les leucémies de type Burkitt.C.R.Acad.Sc.Paris 239,427-429.

Levine E.G.,Arthur D.C.,Frizzera G.,Peterson B.A.,Hurd D.D. and Bloomfield C.D. 1985. They are differences in cytogenetic abnormalities among histologic subtypes of the non-Hodgkin's lymphomas. Blood 66,1414-1422.

Lewis W.H.,Michalopoulos E.E.,Williams D.L.,Minden M.D. and Mak T.W. 1985. Two T-cell leukemia patients with chromosomal translocations have breakpoints within the human T-cell antigen receptor α-chain locus. Nature 317,544-555.

Mathieu-Mahul D.,Caubet J.-F.,Bernheim A.,Mauchauffé M.,Palmer E., Berger R. and Larsen C.J. 1985. Molecular cloning of a DNA fragment from human chromosome 14 (14q11) involved in T-cell malignancies. Embo J. 4,3427-3433.

Mathieu-Mahul D.,Bernheim A.,Sigaux F.,Daniel M.-T.,Larsen C.-J. and Berger R. 1986. Translocation t(1;14) et remaniement du gène de la chaine alpha du récepteur T dans une leucémie aigüe lymphoblastique T. C.R.Acad.Sc.Paris 302,Série III,525-528.

Mathieu-Mahul D.,Chen L.K.,Larsen C.-J. and Bensussan A. 1986. Detection of T cell receptor α chain locus rearrangements in human antigen-specific T lymphocyte clones using a DNA probe specific for chromosome region 14q11. J.Exp.Med.163,1308-1313.

McKeithan T.W.,Shima E.A.,Le Beau M.M.,Minowada J.,Rowley J.D. and Diaz M.O. 1986. Molecular cloning of the breakpoint junction of a human 8;14 translocation involving the T-cell receptor α-chain gene and sequences 3' to c-myc. Proc.Natl.Acad.Sc.USA (in press).

Murphy W.,Sarid J.,Taub R.,Vasicek T.,Battey J.,Lenoir G. and Leder P. 1986. A translocated human c-myc oncogene is altered in a conserved coding sequence. Proc.Nat.Acad.Sci.USA 83,2939 -2943.

Nowell P.C.,Vonderheid E.C.,Besa E.,Hoxie J.A.,Moreau L. and Finan J. 1986. The most common chromosome change in 86 chronic B-cell or T-cell tumors : A 14q32 translocation. Cancer Genet. Cytogenet. 19,219-227.

Pedersen-Bjergaard J.,Andersson P. and Philip P. 1986. Possible pathogenetic significance of specific chromosome abnormalities and activated proto-oncogenes in malignant diseases of man. Scand.J.Haematol. 36,127-137.

Pelicci P.-G.,Knowles D.M.II,Magrath I. and Dalla-Favera R. 1986. Chromosomal breakpoints and structural alterations of the c-myc locus differ in endemic and sporadic forms of Burkitt lymphoma. Proc.Natl.Acad.Sci.USA 83,2984-2988.

Piechaczyk M.,Yang J.Q.,Blanchard J.M.,Jeanteur P. and Marcu K. 1985. Posttranscriptional mechanisms are responsible for accumulation of truncated c-myc RNAs in murine plasma cell tumors.Cell 42,589-597.

Sacchi N.,Watson D.K.,Guerts van Kessel A.H.M.,Hagemeijer A., Patterson D. and Papas T.S. 1986. Hu-ets-1 and Hu-ets-2 genes are transposed in acute leukemias with (4;11) and (8;21) translocations.Science 231,379-382.

Sandberg A.A. 1980. The Chromosomes in Human Cancer and Leukemia. New York : Elsevier.

Shima E.A.,Le Beau M.M.,McKeithan T.W.,Minowada J.,Showe L.C.,Mak T.W.,Minden M.D.,Rowley J.D. and Diaz M.O. 1986. Gene coding the α chain of the T-cell receptor is moved immediately downstream of c-myc in a chromosomal 8;14 translocation in a cell line from a human T-cell leukemia. Proc.Natl.Acad.Sci.USA 83,3439-3443.

Sigaux F.,Berger R.,Bernheim A.,Valensi F.,Daniel M.T. and Flandrin G. 1984. Malignant lymphomas with band 8q24 chromosome abnormality : a morphologic continuum extending from Burkitt's to immunoblastic lymphoma. Br.J.Haematol. 57,393-405.

Tsujimoto Y.,Finger L.R.,Yunis J.,Nowell P.C. and Croce C.M.1984. Cloning of the chromosome breakpoint of neoplastic B cells with the t(14;18) chromosome translocation. Science 226,1097-1099.

Tsujimoto Y.,Cossman J.,Jaffe E. and Croce C.M. 1985. Involvement of the bcl-2 gene in human follicular lymphoma. Science 228, 1440-1143.

Tsujimoto Y.,Gorham J.,Cossman J.,Jaffe E. and Croce C.M. 1985. The t(14;18) chromosome translocations involved in B-cell neoplasms result from mistakes in VDJ joining. Science 229, 1390-1393.

Tsujimoto Y.,Jaffe E.,Cossman J.,Gorham J.,Nowell P.C. and Croce C.M. 1985. Clustering of breakpoints on chromosome 11 in human B-cell neoplasms with the t(11;14) chromosome transloca--tion. Science 315,340-343.

Ueshima Y., Fukuhara S.,Hattori T.,Uchiyama T.,Takatsuki K. and Uchino H. 1981. Chromosome studies in adult T-cell leukemia in Japan : Significance of trisomy 7. Blood 58,420-425.

Williams D.L.,Look A.T.,Melvin S.L.,Roberson P.K.,Dahl G.,Flake T. and Stass S. 1984. New chromosomal translocations correlate with specific immunophenotypes of childhood acute lymphoblastic leukemia. Cell 36,101-109.

Yunis J.J.,Oken M.M.,Theologides A.,Howe R.B. and Kaplan M.E. 1984 Cancer Genet.Cytogenet.13,17-28.

Zech L.,Gahrton G.,Hammarström L.,Juliusson G.,Mellstedt H.,Robèrt K.-H. and Smith C.I.E. 1984. Inversion of chromosome 14 marks human T-cell chronic lymphocytic leukemia.Nature 308,858-860.

Zech L.,Godal T.,Hammarström L.,Mellstedt H.,Smith C.I.E.,Tötter--man T. and Went M. 1986. Specific chromosome markers involved with chronic T lymphocyte tumors. Cancer Genet.Cytogenet. 21, 61-77.

Retroviruses with two oncogenes

D. Stehelin and P. Martin

INSERM U.186, Institut Pasteur, 15 rue C. Guerin, 59019 Lille, France

Less than a decade ago the first retroviral oncogene was discovered (1) and shown to derive from a normal cellular gene (2). Today over twenty different retroviral oncogenes have been found, all derived from cellular counterparts. How a single oncogene can cause cancer is still not fully understood. Natural cancers are known to involve multi-step molecular changes, raising a paradox as compared to cancers induced bu retroviral oncogenes. To turn a cellular gene into an active retroviral oncogene, an overexpression due to the very potent retroviral promotor may suffice in a few cases, whereas in most cases several modifications are probably required. These modifications might involve enhanced transcription, truncation of the gene, specific mutations, deletions or insertions (Fig. 1) as well as the viral route of transmission.

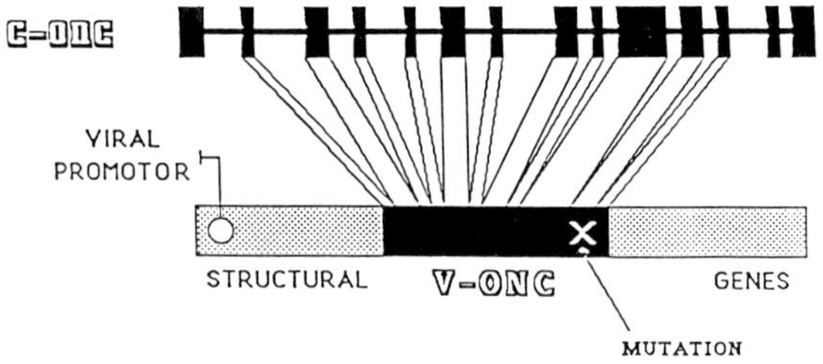

- Transduction of a truncated gene
- Linkage to structural viral genes
- Transcription under a strong viral promotor
- Possible mutations in the transduced gene

Figure 1 : Possible differences between V-onc and C-onc.

In this respect a retroviral oncogene may then be regarded as a single gene having accumulated stepwise, through many replication cycles and selections by researchers (these viruses do not exist in nature) several modifications in order to become a potent cancer gene. Such a process represents a somewhat artificial situation difficult to be challenged for statistical reasons by a cellular oncogene during the life-time of the host-cells. Nevertheless, cellular oncogenes that were identified through retroviruses seem to be involved in some natural cancers, where they are for example de-regulated through chromosomal translocation (c-myc) or mutations[13] (c-ras). Although they appear much "weaker" transforming genes than their viral counterparts, they legitimate the use of retrovirus as tools. It follows that natural cancers are more likely to involve probably single alterations on several distinct genes that cooperate in the tumorigenic process. Recently a first example of cooperation was demonstrated by in-vitro studies showing that the activated ras and myc oncogenes could transform a rat embryo fibroblast in a way that neither gene could achieve alone (3) and some human tumor cell-lines were found to contain the cellular ras and myc oncogenes both altered. How can we identify more such cooperative genes ? Among a few other methods retroviruses may again represent useful tools. In 1979, studying acute avian leukemia retroviruses, we raised the possibility (4) that some retroviruses undergoing modifications could have, among those, transduced not one, but two distinct cellular genes that could cooperate in the viral-proned transformation processes. There are now three examples of such retroviruses (Fig.2) that we would like to review briefly.

Avian erythroblastosis virus (AEV).

The genome of AEV contains the two oncogenes erbA and erbB (5). Deletion mutants in one or the other oncogene have shown that in adult bone marrow the erbB product transforms erythroblasts which subsequently c annot mature properly anymore. The erbA product alone does not modify erythroid cells in a detectable way (6). In contrast, wild type AEV blocks tightly erythroid cells at an immature stage (erythroid colony forming units or CFUe) (7). Thus erbA potentiates the transforming activity of the erbB gene product. The nucleotide sequencing of the erbA gene appears unrelated to any of the oncogenes characterized so far and related to carbonic anhydrases, enzymes known to play a major role in red blood cells, the progenitors of which are precisely the target cells for AEV. Other studies performed on embryonic tissue of the primitive streak indicate that AEV affects also target cells that appear to preceed the BFUe stage.

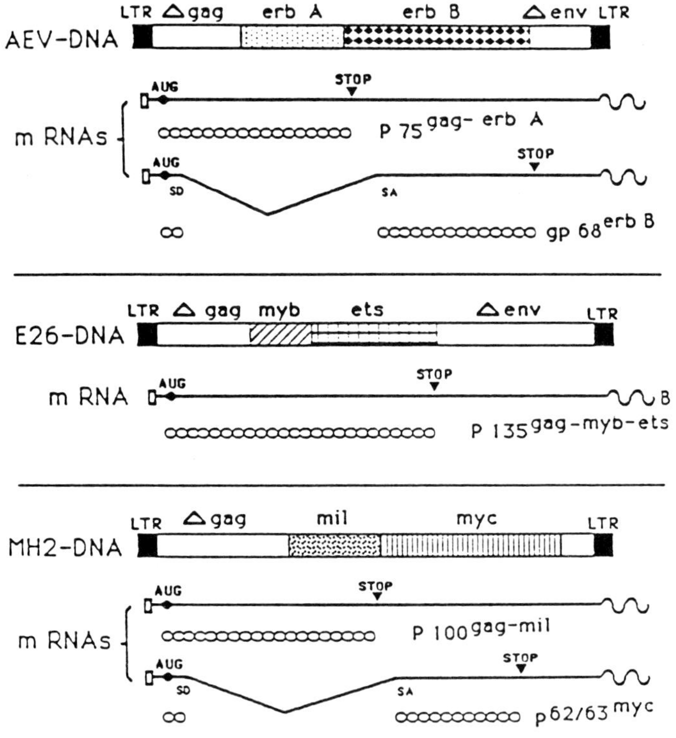

Figure 2: Retroviruses with two oncogenes.

Avian erythroblastosis virus E26

The genome of E26 virus contains the two oncogenes myb and ets (8), expressed in infected cells as a triple fusion protein $P130^{gag-myb-ets}$ (5). E26 appears able to transform uncommitted erythroid-myeloid hemopoietic cells as well as cells commited in the erythroid and myeloid lineages (Moscovici et al).

How the two oncogenes myb and ets are involved in the transforming properties of E26 virus awaits the construction of mutants deleted in one or the other of the genes. We recently obtained a molecular clone of E26 provirus that is biologically active (D. Leprince et al, in preparation) and that should facilitate the construction of such mutants.

Mill-Hill-2 retrovirus (MH2)

The avian retrovirus Mill-Hill-2 (MH2) is a replication defective retrovirus that causes, like other avian myc containing retroviruses (MC29, CMII, OK10), mainly liver and kidney carcinomas in the chicken, and transforms chicken fibroblasts and macrophages in culture (5). MH2 appears more aggressive than the other myc-containing viruses in its tumorigenic potential and its genome contains a second oncogene, mil (9), yielding in infected cells the two oncproteins $P100^{gag-mil}$ and $p62/63^{myc}$ (10). In order to examine the respective roles of the two proteins in the transformation process, we attempted the isolation of spontaneous or constructed mutant expressing properly only the mil or only the myc oncogene product. Two classes of spontaneous mutants were isolated with the group of G. Calothy (Institut du Radium, Orsay, France) using fibroblasts and neuroretinal (RN) cells prepared from 7 day old chick embryos. The choice of this latter cell system was bount to the observation that MH2 wild type virus was shown to stimulate the growth and to transform NR cells that remain usually quiescent in culture (11, and Table I). Such an effect was not detected with retroviruses like MC29 (Table I) that contain only the myc oncogene (11) and transform readily chicken fibroblasts (5). Class I mutants (Table I) were selected on the basis of their inability to induce NR cell proliferation, whereas they still transformed fibroblasts. MH2 Cl 16 exhibited such properties and was shown to suffer an extensive deletion in the mil gene (12). A mutant (MH2-OB) with similar biological properties was constructed by inducing a frameshift in the v-mil gene in a molecularly cloned MH2 provirus, resulting in a premature termination of the v-mil translation product (13).

Virus	Fbl*	NR*		Apparent functional oncogene
	T*	M*	T*	
MC29	+	−	−	myc
wt*MH2	+	+	+	myc + mil
MH2 Cl 16	+	−	−	myc } (Class I)
MH2 OB	+	−	−	myc
MH2 PA 200	−	+	−	mil } (Class II)
MH2 LI 200	−	+	−	mil

* Fbl= fibroblasts; NR= neuroretinal cells;
T= transformed; M= mitogenized; wt = wild type.

Table I: Characterization of MH2 mutants

The class II mutant (i.e. MH2 PA 200) was selected on the ground that it induced NR cells to proliferate with the same efficiency as wt-MH2, but failed to transform fibroblasts (Table I). This mutant also failed to morphologically transform NR cells and showed upon analysis that it was extensively deleted in the myc gene (14). A mutant (MH2 LI 200) with similar biological properties (Table I) was constructed by inducing a frameshift in the v-myc gene of molecularly cloned MH2 provirus that resulted in a premature termination of the v-myc translation product.

The results presented here (13) indicate that the ability to induce sustained proliferation and transformation of NR-cells from 7 day-old chicken embryos is a remarkable property distinguishing MH2 among other myc-containing retroviruses, and requiring the coordinate expression of both mil and myc oncogenes. Class I mutants lacking a functional mil gene (MH2 Cl 16, MH2 OB) do not induce NR cell-proliferation nor transformation, although they still transform fibroblasts (and macrophages : Ref. 15). Conversely, mutants expressing only the mil oncogene (MH2 PA 200, MH2 LI 200) induce NR cell proliferation without morphological transformation.

So far, the viral myc oncogene (or the large T of Polyoma virus) was shown to cooperate with the EJ bladder carcinoma activated ras oncogene (or middle T of Polyoma virus) for rat embryo fibroblasts transformation (3). We show now that the myc oncogene can also cooperate with the mil oncogene (the latter being structurally related to the src-gene family, 16) for the transformation of NR cells. Thus the myc oncogene may cooperate with two distinct types of oncogenes depending on the cell-types considered. Whether myc plays a key role in two distinct pathways leading to transformation, or whether the three types of oncogenes (myc-,ras- and mil-like) belong to a single pathway leading to cell-growth stimulation, where transformation can occur when some of them become deregulated whereas others might be perhaps constitutively expressed at specific stages of cell-maturation or in specific cell-types, remains to be examined. Earlier work on Polyoma might be recalled in this respect. Although the oncogenes of this virus were not shown to have cellular counterparts, the large T and middle T Polyoma genes were shown to cooperate for fibroblast transformation as do the activated myc and ras genes, respectively (17). In addition, the middle T was shown (18) to bind specifically in the transformed fibroblast the cellular protein $pp60^{src}$ that was shown activated in the complex.

In conclusion research begins slowly to unwrap the cooperation of genes in normal and pathological cell-growth stimulation, and retroviruses with double oncogenes represent convenient tools in such investigations. They have allowed to discover three new oncogenes (erbA, ets and mil) that potentiate previously described oncogenes (erbB, myb and myc, respectively) in their transforming activity or allow these viruses to transform new target cells. Wether the corresponding couples of cellular oncogenes participate in the formation of natural human cancers is under investigation.

REFERENCES

01 - Stéhelin,D., Guntaka,R.V., Varmus,H.E. and Bishop,J.M. (1976) J. Mol. Biol. 101 : 349-365

02 - Stéhelin,D., Varmus,H.E., Bishop,J.M. and Vogt,P.K. (1976) Nature 260 : 170-173

03 - Land,H., Parada,L.F. and Weinberg,R.A. (1983) Nature 304 : 596-602

04 - Roussel,M. et al. (1979) Nature 281 : 425-455

05 - Graf,T. and Stéhelin,D. (1982) Biophys Biochim. Acta 651 : 245-271

06 - Frykberg,L. et al. (1983) Cell 32 : 227-238

07 - Samarut, J. and Gazzolo, L. (1982)
Cell 28 : 921-929
08 - Leprince,D. et al. (1983)
Nature 306 : 395-397
09 - Coll, J. et al. (1983)
Embo J. 2 : 2189-2194
10 - Pachl, C., Biegalke,B. and Linial,M. (1983)
J. Virol 45 : 133-139
11 - Calothy,G. et al. (1979)
Cold Spring Harbor Symp. Quant. Biol. 44 : 983-990
12 - Martin,P. et al. (1985)
J. Virol. 57 : 1191-1194
13 - Bechade et al. (1985)
Nature 316 : 559-562
14 - Martin et al. (1986)
Virology (in press)
15 - Graf et al. (1986)
Cell 45 : 357-364
16 - Galibert,F. et al. (1984)
Embo J. 3 : 1333-1338
17 - Cuzin,F. (1984)
Biophys. Biochim. Acta 781 : 193-204
18 - Courtneidge,S.A. and Smith,A.E. (1983).
Nature 303 : 435-439

Ph^1 chromosome: cytogenetics and molecular aspects

A. Hagemeijer and A. de Klein

Department of Cell Biology and Genetics, Erasmus University, P.O. Box 1738, 3000 DR Rotterdam, The Netherlands

INTRODUCTION

In 1960, Nowell and Hungerford described for the first time an abnormal small chromosome, the Philadelphia chromosome (Ph^1) in the leukemic cells of two patients with chronic myeloid leukemia (CML). Use of banding techniques showed that this Ph^1 originated from a translocation of the distal part of chromosome 22 to the long arm of chromosome 9: t(9;22)(q34;q11) (Rowley

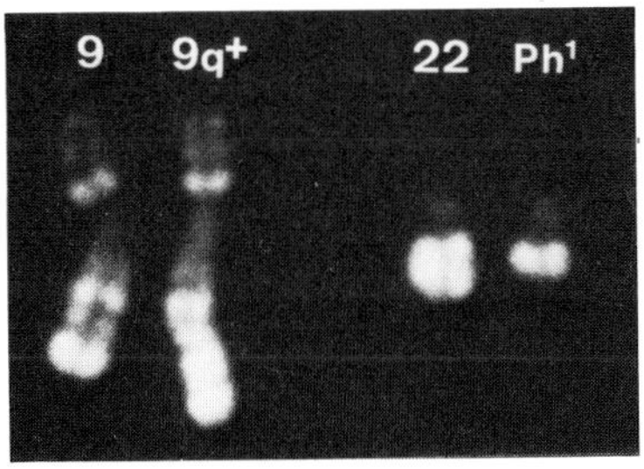

Fig. 1. Standard Ph^1 translocation: t(9;22)(q34;q11) in CML using R-bands

1973). Cytogenetic analysis of bone marrow cells of patients with CML and other hematopoietic diseases revealed that the Ph^1 chromosome is consistently found in more than 96% of the CML cases, usually as a result of the standard t(9;22); although cytogenetic variants of the Ph^1 translocation have been reported in about 3-8% of the Ph^1(+) CML cases. A small percentage (3-7%) of CML cases are without a Ph^1 chromosome (Ph^1 negative).

Furthermore the Ph^1 chromosome has been reported in other hematopoietic disorders, particularly in the adult type of acute lymphoblastic leukemia (ALL) (Rowley 1980, Sandberg 1980). The discovery of cellular oncogenes and their localization near the breakpoints in several specific chromosomal rearrangements in various forms of cancer suggest a key role of these genes in the development of these neoplasia. Of interest in the Ph^1 translocation are the cellular oncogenes c-abl and c-sis which have been assigned to chromosome 9 and 22 respectively. The involvement of these oncogenes in the specific Ph^1 translocation has been studied using molecular and cytogenetic techniques. These studies have provided an important and fundamental insight in the nature of the Ph^1 chromosome in CML and other Ph^1 positive hematopoietic diseases.

ONCOGENES INVOLVED IN THE PH^1 TRANSLOCATION

Recently a family of related genes, termed oncogenes or proto-oncogenes has been identified (Varmus 1984). These genes are the normal cellular homologues of the transforming genes of acutely transforming retroviruses. The human c-abl sequences, homologous to the transforming gene (v-abl) of Abelson murine leukemia virus, have been cloned and analysed (Heisterkamp et al. 1983a, Shtivelman et al. 1985). These c-abl coding regions are distributed over 50 kb of genomic DNA and are dispersed over at least 11 exons. The precise 5' end of the human c-abl gene still remains to be determined. The human c-abl sequences have been localized to the long arm of chromosome 9 (Heisterkamp et al. 1982). To study the localization of c-abl in the Ph^1 translocation, Southern blot analysis of a panel of somatic cell hybrids was performed. The somatic hybrids were obtained by fusion of rodent cells with $Ph^1(+)$ leukemic cells of CML patients (Geurts van Kessel et al. 1981).

Analysis of DNA from hybrids containing either the 9q+ or the 22q- chromosome demonstrated the translocation of human c-abl from chromosome 9 to the Ph[1] (De Klein et al.1982). c-sis represents the cellular homologue of the transforming gene (v-sis) of Simian sarcoma virus The human c-sis gene is localized on chromosome 22 (Dalla-Favera et al. 1982). Hybridization of a c-sis probe to DNA of somatic cell hybrids revealed that this gene is translocated to the 9q+ derivative in the Ph[1] translocation (Groffen et al. 1983).

MOLECULAR CLONING OF THE BREAKPOINTS

The breakpoint on chromosome 9 is located on the very telomeric band 9q34, and the consistent translocation of a relatively small fragment of less than 5000 kb containing v-abl homologous sequences suggested a localization near or within the c-abl locus. Southern blot analysis and subsequent cloning and characterization of chimeric breakpoint fragments have demonstrated the localization of chromosome 9 breakpoints adjacent to the c-abl gene (Heisterkamp et al. 1983b; Grosveld et al. 1986). These breakpoints are all upstream (5') of the v-abl homologous sequences at variable distances from 14 up to more than 100 kb. The recent identification of c-abl exons more 5' (Shtivelman et al. 1985; Grosveld et al. 1986) indicate that some of these breakpoints fall within intron sequences of the c-abl gene whereas other breakpoints, among them the chromosome 9 breakpoint of the CML-derived cellline K562, are located 5' of the presumable first c-abl exon. Restriction enzyme analysis of DNA from over 50 CML patients revealed that the breakpoints on chromosome 22q- clustered all within the 5.0 kb Bgl II fragment that was called bcr for breakpoint cluster region (Groffen et al, 1984). The 3' site of this bcr usually detects the 9q+ chimeric fragment whereas the

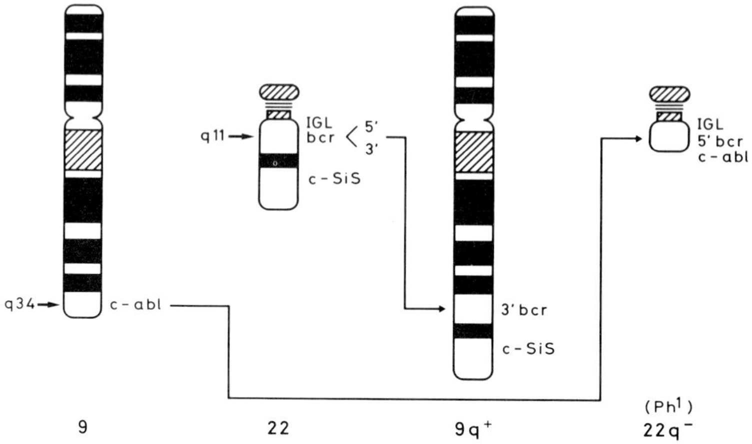

Fig. 2. Schematic representation of the Ph[1] translocation indicating the regional localization of the *bcr*, c-*abl*, c-*sis* and *IGL* genes.

5'site of this *bcr* detects the 22q- chimeric fragment (Fig. 2). Using chromosome 22 sequences as probes, the normal *bcr* region was cloned and found to the part of a gene of unknown function on chromosome 22, and that contains at least 18 exons (Heisterkamp et al. 1985). The 5.0 kb Bgl II fragment previously identified is centrally situated and contains 4 small exons (b1-b4). Mapping experiments and sequencing of several cloned Ph[1]-breakpoint fragments from different patients indicated that breakpoints occur in introns between *bcr* exons b2 and b3 or exons b3 and b4. At this point two features characterize the Ph[1] chromosome in Ph[1](+) CML causing the standard t(9;22): 1) translocation of c-*abl* and 2) a breakpoint in the *bcr* region as defined by restriction enzyme analysis (Fig. 2).

ANALYSIS OF CYTOGENETIC AND CLINICAL VARIANTS USING SOUTHERN BLOT AND IN SITU HYBRIDIZATION TECHNIQUES

The results described above provided new tools, i.e. probes specific for the oncogenes c-*abl* and c-*sis*, and probes hybridizing with various parts of the *bcr* region, for the further characterization of the Ph^1 chromosome in leukemic cells of patients with CML or other hematopoietic disorders.

A. Variants of Ph^1 translocation

Three forms of cytogenetic variants have been reported: firstly,simple translocations between chromosome 22 and another chromosome, without visible involvement of chromosome 9,secondly, complex translocations involving 9, 22 and at least one other chromosome (in this case, the third chromosome is recipient of the deleted part of chromosome 22q- while 9 is recipient of the deleted part of the third chromosome); thirdly, a rare type of variant called "masked Ph^1", where the Ph^1 is the recipient of a part of another chromosome. Using in situ hybridization technique and in some cases Southern blot we analysed 5 cases of complex variants, 7 cases of apparently simple variants and 4 cases of masked Ph^1. The results of these studies clearly showed that c-*abl* oncogene was consistently translocated to the Ph^1, also in so-called simple translocation variants and one masked Ph^1 where chromosome 9 was not visibly altered (Fig. 3). Here also a breakpoint on chromosome

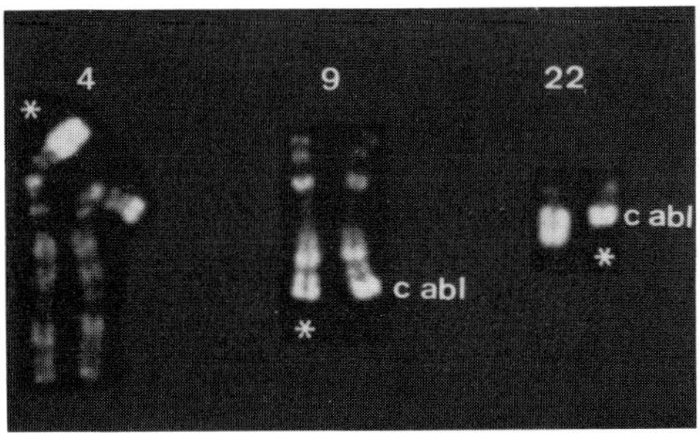

Fig. 3. Variant of Ph^1: t(4;9;22). *identifies the derivative chromosomes. The hybridization sites of c-*abl* probes are indicated.

22q- mapped in the bcr and as a consequence the c-sis oncogene was found to be located on the derivative chromosome carrying the distal part of chromosome 22 (Bartram et al. 1983, 1984, Hagemeijer et al. 1984, 1985 and unpublished observations).

B. Ph^1(-) CML

A small number of CML patients are without a Ph^1 chromosome, although in some cases translocations were observed, but with an apparently intact chromosome 22. In Rotterdam we studied 7 cases of Ph^1(-) CML and Morris et al. (1986) reported on 5 cases of Ph^1(-) CML with a normal karyotype. No translocations of c-abl or bcr breakpoints were observed in 5 of our cases with a normal karyotype and in 3 of the cases studied by Morris et al. (1986). In contrast, the 2 cases with translocations apparently not involving chromosome 22, i.e. t(9;12) and t(9;13;15) a breakpoint in bcr was detected by Southern blot, although c-sis oncogene was not translocated. Extensive in situ hybridization studies of the patient with t(9;12)(q34;q21) demonstrated the juxtaposition of 5'bcr and c-abl on chromosome 12q- (Bartram et al, 1985). Similarly in patient 1 from Morris et al, (1986), c-abl was found to be inserted within the bcr-region of chromosome 22 by in situ hybridization. Clinically, the patients with bcr/c-abl rearrangement showed a history very similar to Ph^1(+) CML cases. In contrast, the true Ph^1(-) cases had the expected shortest survival and presented features reminding of myelodysplastic syndromes.

C. Ph^1(+) ALL AND AML

A Ph^1 translocation, cytogenetically undistinguishable of the Ph^1 in CML is found not unfrequently in acute lymphocytic leukemia (ALL). Our own studies using Southern blot analysis showed an heterogeneous picture: 6 ALL patients showed bcr rearrangements as observed in CML; in 3 other patients recombination involving 5'bcr sequences could be demonstrated but the corresponding 3'bcr sequences were not detectable and 5 patients did not show any bcr rearrangements at all (De Klein et al. 1986, submitted). Similar results were obtained by Rodenhuis et al (1985) and Erikson et al (1986). Using in situ hybridization it was demonstrated that c-abl was translocated to the Ph^1 chromosome even in cases without a bcr rearrangement (De Klein et al, submitted). In these latter cases the 22q11 breakpoint is located proximal to the bcr region (Erikson et al, 1986) probably distal to the immunoglobulin gene locus $C\lambda$, although in one case in situ hybridization studies suggested a breakpoint within the $C\lambda$ region. In AML the Ph^1 translocation is found in a small number of cases. Studies of such cases revealed an heterogeneity regarding the bcr breakpoint similar to our findings in ALL (Bartram, personal communication). The heterogeneity in bcr rearrangements observed in Ph^1(+) acute leukemia is in contrast with the consistent results obtained in Ph^1(+) CML. This suggests that, if c-abl is inappropriately expressed in Ph^1(+) acute leukemia cells without bcr rearrangement, the genetic mechanism of activation must be different from that reported in CML.

TRANSCRIPTION OF C-ABL IN CML

Gale and Canaani (1984) investigating the expression of c-abl in Ph^1(+) CML patients and various controls found in addition to the normal transcripts of 6 or 7 kb a new RNA transcript of about 8.5 kb. This aberrant c-abl mRNA strictly correlated with Ph^1(+) CML and was not observed in the various Ph^1(-) healthy or leukemic controls. Similar observations were made when we studied the expression of the bcr gene in Ph^1(+) CML patients (Stam et al, 1985): in these Ph(+) leukemic cells an extra 8.5 kb bcr specific mRNA was present besides the normal 7.5, 7.0 and 4.5 kb transcripts.

This 8.5 kb mRNA hybridized only to a 5'bcr specific probe and was not detectable with a more 3'bcr probe. Since the 5'bcr and c-abl gene as a result of the Ph[1] translocation are juxtaposed in the same transcriptional orientation, these results suggested that the 8.5 kb mRNA is a chimeric molecule that contains in addition to c-abl sequences the code of the 5' region of the bcr gene.

CONSEQUENCES OF GENETIC RECOMBINATIONS IN PH[1]

The strong correlation between the bcr/c-abl rearrangement and CML was further underlined by the cloning of a 22q- breakpoint from the cellline K562. Although this celline was established from Ph(+) cells of a CML blastcrisis it does not contain a cytogenetically detectable Ph[1] chromosome (unpublished observations). The cloning of the 22q- breakpoint region from this cellline revealed that the entire breakpoint region including the 5'bcr part and the c-abl gene was amplified (Grosveld et al. 1986). Since also the more proximal Cλ locus on chromosome 22 is amplified this cellline probably contains amplified remnants of the original Ph[1] chromosome (Collins et al. 1984). Both the genes bcr and c-abl are fused in the same transcriptional orientation but are located at least 100 kb apart of each other. Nevertheless, this cellline shows an enhanced expression of the 5'bcr and 3'c-abl specific mRNA of 8.5 kb (Grosveld et al. 1986). Partial cDNA's for the 8.5 kb mRNA molecules were cloned from K562 (and EM.2) c-DNA libraries independently by Shtivelman et al (1985) and Grosveld et al. (1986). Restriction enzyme analysis and DNA sequencing demonstrated that in K562 the exon b3 of the bcr gene is adjacent to the c-abl exon a1. Translation of this chimeric cDNA into protein shows one open reading frame that is compatible with both the predicted bcr and the known c-abl reading frames. This constituted a direct proof that the 8.5 kb mRNA is a chimeric molecule that can code for a bcr/c-abl fusion protein. Dependent on the location of the 22q- breakpoints either exon b2 or exon b3 of bcr can be spliced to c-abl exon a1 since both exons b2 and b3 have the 3' splice donor site after the first nucleo-tide of the codon. Both would result in a translatable mRNA of about 8.5 kb. A difference of 75 bp (the size of exon b3) is beyond the detection limit. This splicing pattern would also explain why the same 8.5 kb mRNA is detected in different patients (Gale & Canaani 1984, Stam et al. 1985) although the distance between the Ph^1 breakpoint and v-abl hybridizing sequences varies from 14 to more than 100 kb.

The implications of the translocation of c-abl sequences to the 3' side of bcr on the Ph chromosome are as follows: transcription is likely to be initiated from the bcr promoter and probably stops at the 3' end of the c-abl gene. By splicing of the precursor RNA the 8.5 kb mRNA is produced in which the versatility of the splicing system accomodates for the large variation in intron size linking bcr and c-abl in different patients. A likely candidate for the translation product of the 8.5 kb mRNA is the abnormally sized 210 kD c-abl protein found in K562 cells which in contrast to normal c-abl has in vitro tyrosine kinase activity (Konopka et al. 1984), very similar to the v-abl gene product (Davids et al. 1985). The bcr moiety of this hybrid molecule could unmask the c-abl tyrosine kinase activity, raising the question whether bcr is sufficient or necessary for this effect. The bcr/c-abl hybrid protein may have transforming activity, but it remains to be established if it plays an essential role in the generation and or maintenance of CML.

REFERENCES

Bartram, C.R., De Klein, A., Hagemeijer, A., van Agthoven, T., Geurts van Kessel, A., Bootsma, D., Grosveld, G., Davies, T., Stone, M., Ferguson-Smith, M.A., Heisterkamp, N., Groffen, J. and Stephenson, J.R. 1983. Translocation of the human c-abl oncogene occurs in variant Ph1-positive but not Ph1-negative chronic myelocytic leukaemia. Nature 306, 277-280.

Bartram, C.R., De Klein, A., Hagemeijer, A., Grosveld, G., Heisterkamp, N. and Groffen, J. 1984. Localization of the human c-sis oncogene in Ph1-positive and Ph1-negative chronic myelocytic leukemia by in situ hybridization. Blood 63, 223-225.

Bartram, C.R., Kleihauer, E., De Klein, A., Grosveld, G., Teyssier, J., Heisterkamp, N. and Groffen, J. 1985. c-abl and bcr rearranged in a Ph-negative CML patient. EMBO J. 4, 683-686.

Collins, S.J., Kubonishi, I., Miyoshi, I., Groudine, M.T. 1984. Altered transcription of the c-abl oncogene in K562 and other chronic myelogenous leukemia cells. Science 225, 72-74.

Dalla-Favera, R., Gallo, R.C., Giallongo, A. and Croce, C.M. 1982. Chromosomal localization of the human homologue (c-sis) of the simian sarcoma virus onc gene. Science 218, 686-688.

Davis, C.R., Konopka, J.B. and Witte, O.N. 1985. Activation of the c-abl oncogene by viral transduction of chromosomal translocation generates altered c-abl proteins with similar in vitro kinase properties. Mol. Cell. Biol. 5, 204-213.

De Klein, A., Geurts van Kessel, A., Grosveld, G., Bartram, C.R., Hagemeijer, A., Bootsma, D., Spurr, N.K., Heisterkamp, N., Groffen, J. and Stephenson, J.R. 1982. A cellular oncogene is translocated to the Philadelphia chromosome in chronic myelocytic leukaemia. Nature 300, 765-767.

Erikson, J., Griffin, C.A., Abbas ar-Rushdi, Valtieri, M., Hoxie, J., Finan, J., Emanuel, B.S., Rovera, G., Nowell, P.C. and Croce, C.M. 1986. Heterogeneity of chromosome 22 breakpoint in Philadelphia positive (Ph+) acute lymphocytic leukemia. Proc. Nat. Acad. Sci. USA 83, 1807-1811.

Gale, R.P. and Canaani, E. 1984. An 8-kilobase abl RNA transcript in chronic myelogenous leukemia. Proc. Nat. Acad. Sc. USA 81, 5648-5652.

Geurts van Kessel, A.H.M., ten Brinke, H., Boere, W.A.M., den Boer, W.C., de Groot, P.G., Hagemeijer, A., Meera Khan, P. and Pearson, P.L. 1981. Characterization of the Philadelphia chromosome by gene mapping. Cytogenet. Cell Genet. 30, 83-91.

Grosveld, G., Verwoerd, T., van Agthoven, T., De Klein, A., Ramachandran, K., Heisterkamp, N., Stam, K. 1986. The chronic myelocytic cell line K562 contains a breakpoint in bcr and produces a chimeric bcr/c-abl transcript. Mol.Cell. Biol. 6, 607-616.

Groffen, J., Heisterkamp, N., Stephenson, J.R., Geurts van Kessel, A., De Klein, A., Grosveld, G. and Bootsma, D. 1983. c-sis is translocated from chromosome 22 to chromosome 9 in chronic myelocytic leukemia. J. Exp. Med. 158, 9-15.

Groffen, J., Stephenson, J.R., Heisterkamp, N., De Klein, A., Bartram, C.R. and Grosveld, G. 1984. Philadelphia chromosomal breakpoints are clustered within a limited region, bcr, on chromosome 22. Cell 36, 93-99.

Hagemeijer, A., Bartram, C.R., Smit, E.M.E., van Agthoven, A.J. and Bootsma, D. 1984. Is the chromosomal region 9q34 always involved in variants of the Phl translocation? Cancer Genet. Cytogenet. 13, 1-16.

Hagemeijer, A., De Klein, A., Godde-Salz, E., Turc-Carel, C., Smit, E., van Agthoven, A. and Grosveld, G. 1985. Translocation of c-abl to masked Ph in chronic myeloid leukemia. Cancer Genet Cytogenet. 18, 95-104.

Heisterkamp, N., Groffen, J., Stephenson, J.R., Spurr, N.K., Goodfellow, P.N., Solomon, E., Carrit, B. and Bodmer, W.F. 1982. Chromosomal localization of human cellular homologues of two viral oncogenes. Nature 299, 747-749.

Heisterkamp, N., Groffen, J. and Stephenson, J.R. 1983a. The human v-abl homologue. J. Mol. Appl. Genet. 2, 57-69.

Heisterkamp, N., Stephenson, J.R., Groffen, J.R., Hansen, P.F., De Klein, A., Bartram, C.R. and Grosveld, G. 1983b. Localization of the c-abl oncogene adjacent to a translocation breakpoint in chronic myelocytic leukemia. Nature 306, 239-242.

Heisterkamp, N., Stam, K., Groffen, J., De Klein, A., Grosveld, G. 1985. Structural organization of the bcr gene and its role in the Phl translocation. Nature 315, 758-761.

Konopka, J.B., Watanabe, S.M., Witte, O.N. 1984. An alteration of human c-abl protein in K562 leukemia cells unmasks associated tyrosine kinase activity. Cell 37, 1035-1042.

Morris, C.M., Reeve, A.E., Fitzgerald, P.H., Hollings, P.E., Beard, M.E.J. and Heaton, D.C. 1986. Nature 320, 281-283.

Nowell, P.C., and Hungerford, D.A. 1960. A minute chromosome in human chronic granulocytic leukemia. Science 132, 1497.

Rodenhuis, S., Smets, L.A., Slater, R.M., Behrendt, H., Veerman, A. 1985. Distinghuishing the Philadelphia chromosome of acute lymphoblastic leukemia from its counterpart in chronic myelogenous leukemia. N. Engl. J. Med. 313, 51-52.

Rowley, J.D. 1973. A new consistent chromosomal abnormality in chronic myelogenous leukaemia identified by quinacrine fluorescence and Giemsa staining. Nature 243, 290-293.

Rowley, J.D. 1980. Phl positive leukaemia, including chronic myelogenous leukaemia. Clin. Haematol. 9, 55-58.

Sandberg, A.A. 1980. The cytogenetics of chronic myelocystic leukemia: Chronic phase and blastic crisis. Cancer Genet. Cytogenet. 1, 217-228.

Shtivelman, E., Lifshitz, B., Gale, R.B. and Canaani, E. 1985. Fused transcript of abl and bcr genes in chronic myelogenous leukemia. Nature 315, 550-554.

Stam, K., Heisterkamp, N., Grosveld, G., de Klein, A., Verma, R.S., Coleman, M., Dosik, H. and Groffen, J. 1985. Evidence of a new chimeric bcr/c-abl in RNA in patients with chronic myelocytic leukemia and the Philadelphia chromosome. N. Engl .J. Med. 313, 1429-1433.

Varmus, H.E. 1984. The molecular genetics of cellular oncogenes. Ann. Rev. Genet. 18, 553-612.

SEX CHROMOSOMES

Conservation *in toto* of the mammalian X-linkage group as a frozen accident

S. Ohno

Beckman Research Institute of the City of Hope Duarte, California 91010, USA

It now seems that the X-linkage group has indeed been conserved *in toto* throughout evolution of not only placental mammals but also of marsupials as well (Ohno et al.,1964a). The unique dosage compensation mechanism that relies on inactivation of one or the other X in females (Lyon,1961) has often been invoked as the cause of this extreme evolutionary conservation. I see no merit in this widely held belief. There apparently exists no dosage compensation mechanism for avian Z-linked genes. In fact, avian species seem to expolit dosage effects of their Z-linked genes for the manifestation of sexual dimorphism. I had a personal experience with a flock of pigeons. A single dose of a particular Z-linked mutant gene in hemizygous ZW females merely diluted the plumage color to creamy yellow, whereas double dose of the same gene in homozygous males made them almost white. In fact, hemizyous females were phenotypic eqivalents of heterozygous males. Similarly in the barred Plymouth Rock breed of chickens, roosters appear considerably lighter in color than hens, for the double dose of Z-linked *Barred* caused white bands in individual feathers to be much wider than white bands of hemizygous females. Yet the Z-linkage group of birds has been conserved as dilligently as the mammalian X-linkage group (Ohno et al,1964b, Baverstock et al,1982).

We thus come to the question of why the particular chromosome was chosen as the sex chromosome in a particular lineage. Quite clearly, there exists no *a priori* requirement to be a sex chromosome, for the mammalian X-chromosome and the avain Z-chromosomes are very different in their genetic contents (Baverstock et al,1982). In fact, the chromosomal sex determining mechanism at its inception apparently try a few randomly chosen pairs of chromosomes as prospective sex chromosomes. Remembering that the X and the Y or the Z and the W at their inceptions are still homomorphic and largely homologous with each other (Ohno,1967), the following findings on various populations and species of frogs of the family *Ranidae* by Wright et al.(1983) appears particularly relevant. In many, but not all, populations of the leopard frog (*Rana pipiens*), the sex determining locus is in rather tight linkage with the two enzyme loci for super oxide dismutase-1 and peptidase C; the Y chromosome carrying a particular pair of alleles of these two enzyme loci, whereas the X-chromosome bearing alternative alleles of the two. In another species, *Rana sphenocephala*, the sex determining locus is now linked to the entirely different set of enzyme loci; phosphoglucomutase-1, fructose 1, 6 diphosphatase and alcohol dehydrogenase 2 (Wright et al.,1983). These apparently random choices indicated that the number of genes involved in sex determination and sexual development is rather small, and that only smaller still of those have to be located on sex chromosomes.

SEX DETERMINING GENES ON THE X CHROMOSOME

The list of X-linked genes has been becoming more voluminous every year. Yet, the fact remains that a great majority of them have nothing whatsoever to do with the sex determination and sexual development. This is exactly what we expected from the foregoing discussion. Glucose-6-phosphate dehydrogenase and phosphoglycerate kinase are sugar metabolizing enzymes, and these sugar metabolizing enzymes have already acquired the inviolable functional competence before the division of prokaryotes from eukaryotes nearly 2 billion years ago. There are genes for purine scavenging and nucleotide synthesis, hypoxanthine-guanine phosphorybosyl transferase and phosphorybosylpyrophosphate synthetase, and then there are genes for lysosomal degradation and detoxification; α-galactosidase A and monoamine oxidase A. Colorblindness genes propably have something to do with retinal pigments and then there are two components of the blood clotting cascade; antihemophilic factor VIII and IX. The former is probably one of the more ridiculous genes in our genome. The gene made of 26 exons encompasses 1.86 X 10^5 base pairs of DNA representing 0.1% of the X-chromosome DNA, one round of transcription of this gene is thought to require more than three hours (Gitschier et al.,1984). When 2,332-amino acid residue long factor VIII is cleaved by thrombin (one of the serine proteases), two active fragments are produced. These active fragments show homology to serum copper-binding protein, ceruloplasmin, on one hand, and too, of all things, discoidin lectins from the slime mold on the other. The more immediate common ancestry, however, is shared with another factor involved in the blood coagulation cascade, factor V. It follows that the genetic content of the mammalian X-chromosome is no different from that of any other chromosome in that it is a true hodgepodge containing parts, but never the whole, of diverse gene sets involved in divergent catabolic and anabolic pathways and cascades.

What are of the genes involved in the sex determination and sexual development? The female being the natural sex in mammals, the *testicular feminization* locus encoding the one and only species of nuclear-cytosol androgen receptor protein can be said to be the true master regulator of extragonadal masculine development, for every androgen-induced developmental response is mediated by this receptor. Accordingly, in its mutational absence, XY individuals with androgen producing testes manifest feminine phenotype. Yet, there seems to be no *a priori* reason for this receptor gene to be on the X chromosome, since this gene is expressed in males and females alike, thus, administration of androgen to genetic females inducing all the extragonadal masculine development. Were it on an autosome, I am confident that the *Tfm* locus would have carried out its assigned function equally well. Indeed, most, if not all, of the enzyme loci involved in synthesis of masculine and feminine steroids reside on autosomes; e.g., that involved in conversion of 17α-hydroxyprogesterone to androstendione and the microsomal aromatase that converts androtendione to estrone and testosterone to estradiol.

The above leaves us with only one other X-linked gene

or gene cluster, this time, involved in the gonadal sex determination. The cause of fertile XY females prevalent in populations of the wood lemming, *Myopus schisticolor* is not to be found on a defective Y chromosome but the femaleness is caused by an internal rearrangement affecting the short arm of the X chromosome (Fredga et al.,1976, Gropp et al.,1976). If the X-linked gene, encoding the plasma membrane receptor for the Y encoded testis determining protein (H-Y antigen?) has been deleted by this rearrangement, the normal Y chromosome might as well be not there. Whatever, the function of this gene locus or gene cluster, is there *a priori* reason for this locus to be on the X-chromosome? It seems to me the reasoning developed with regard to the location of the *Tfm* locus applies equally well here. Inasmuch as only genetic males with the normal Y chromosome can express the all powerful testis organizing protein, the receptor to it might as well be expressed in males and females alike. Accordingly, there seems to be no harm if this locus too resides on an autosome. In fact, it would be recalled that a great majority of the XY/XX blastocyt fusion chimeras develop as males which reveals that in mosaic indifferent gonads, XY cells are capable of enticing neighboring XX cells to participate in testiscular organogenesis (McLaren,1976). Such enticement would not be possible unless XX gonadal cells too express the receptor for the Y-derived testis determining protein. The conclusion to be drawn from the above seems to be that there is no gene on the mammalian X chromosome that has *a priori* reason to be there. This applies even to a few genes that are intimately involved in the sex determination and sexual development.

WHY THE CONSERVATION?

We are thus led to the conclusion that the X chromosome is the X chromosome merely because it is a pairing partner of the Y chromosome during male meiosis, being the X chromosome having no intrincic significance by itself. It is granted that the above offers no answer to the very stringent evolutionary conservation *in toto* of the mammalian X-linkage group. Before considering the proper answer, however, I should point out that conservation and not innovation has always been the rule of evolution ever since the beginning of life on this earth more than 4 billion years ago. Thus, one should never be surprised by evolutionary conservation. I have not heard of any means which permit us to calculate the intrincic rate of linkage group rearrangements in evolution. But one can readily do this with a reasonable certainty as to the fate of individual genes. Even if totally ignored by natural selection as pseudogenes have to be, the base sequence of vertebrates in general and mammals in particular would accumulate mutational base substitutions only at the rate of 10^{-9}/base pair/year. At this rate, the protein would undergo 1% amino acid sequence change only every one million years. Not surprisingly, the halflife of the enzyme loci that became redundant due to duplication has been calculated as 50 million years. In essence, the gene that became useless may continue to encode a functional enzyme for 50 million years or more (Ohno,1985). If the similar rate applies to linakge group rearrangements, a given linkage group may have been conserved in *in toto* for 50 million years for no reason at all. It would thus appear that the observed evolutionary conservation *in toto* of the mammalian X-linkage group for 120 million years or thereabout, needs not invoke a very strong selection against break ups, for at least half of the reason for conservation can be found in the inherent evolutionary tendency for conservation of all the past accomplishments.

What has been this little added incentive that contributed to the *in toto* linkage group conservation for 120 million years instead of 50 million years or so? Inasmuch as the so called XY_1Y_2/XX scheme of sex determining mechanism has been found in a considerable number of divergent mammalian species, it is clear that so long as the break point on the X chromosome is very near to its one or the other terminus, X-autosome translocations are quite compatible with speciation and adaptive radiation. But all X-autosome translocations that broke the X chromosome into two pieces, each of substantial size, have been found to cause male sterility. This seems to be that needed little incentive that caused the mammalian X-linkage group to be conserved *in toto* for 120 million years, or so. Cattanach's X-autosome translocation of the mouse appears as an apparent exception to this rule of male sterility. However, it should be pointed out that although the X chromosome has indeed been broken down to two substantial pieces, they have subsequently reunited after the insertion of an autosomal piece from Linkage group 7(Ohno and Cattanach,1962).

SUMMARY

It appears that the chromosomal sex determining mechanism at its inception more or less randomly choose a pair of chromosomes for the future X and Y or the Z and W. Although both have been equally conserved in evolution, the avian Z-chromosome and the mammalian X-chromosome are very different in their genetic contents and the dosage compensation mechanism operates only in the latter. As to the genetic content of the mammalian X-chromosome, it is no different from any other chromosomes being hodgepodge of diverse genes representing parts but never the whole of gene sets involved in various catabolic and anabolic pathways and cascades. Although there are small number of X-linked genes that are directly involved in the sex determination and sexual development, there is no *a priori* reason for them to be X-linked for their autosomal residence would not have interfered with their assigned functions. I come to the conclusion that the X-chromosome is the X-chromosome merely because it is a pairing partner of the Y chromosome during male meiosis. The evolutionary conservation *in toto* of the mammalian X-linakge group for 120 million years or thereabout appears to have been due largely to the general evolutionary tendency for conservation of all the past accomplishment. The added incentive to the above is the male sterility caused by most X-autosome translocations.

REFERENCES

Baverstock, P.R., Adams, M., Polkinghorne, R.W. and Gelder, M. 1982. A sex-linked enzyme in birds:

REFERENCES (continued)

Z-chromosome conservation and lack of dosage compensation. *Nature* 296, 763-766.

Fredga, K., Gropp, A., Winking, H. and Frank, F. 1976. Fertile XX- and XY-type females in the wood lemming (*Myopus schisticolor*). *Nature* 261, 255-257.

Gitschier, J., Wood, W.I., Goralka, T.M., Wion, K.L., Chen, E.Y., Eaton, D.H., Vehar, G.A., Capon, D.J. and Lawn, R.M. 1984. Characterization of human factor VIII gene. *Nature* 312, 326-330.

Gropp, A., Winking, H., Frank, F., Noack, G. and Fredga, K. 1976. Sex-chromosome aberrations in wood lemmings (*Myopus schisticolor*). *Cytogenet. Cell Genet.* 17, 343-358.

Lyon, M.F. 1961. Gene action in the X-chromosome of the mouse (*Mus musculus L.*). *Nature* 190, 372-373.

McLaren, A. 1976. *Mammalian Chimaeras, Development and Cell Biology 4*. London: Cambridge Univ. Press.

Ohno, S. and Cattanach, B.M. 1962. Cytological study of X-autosome translocation in *Mus musculus*. *Cytogenetics* 1, 129-140.

Ohno, S., Becak, W. and Becak, M.L. 1964a. X-autosome ratio and the behavior pattern of individual X-chromosomes in placental mammals. *Chromosoma* 15, 14-30.

Ohno, S., Stenius, C., Christian, L.C., Becak, W. and Becak, M.L. 1964b. Chromosomal uniformity in the avian subclass *Carinatae*. *Chromosoma* (Berl.) 15, 280-288.

Ohno, S. 1967. *Sex Chromosomes and Sex-linked Genes*. Vol. 1, Monographs on endocrinology. A. Labhart, T. Mann, L.T. Samuels, eds., Heidelberg-Berlin-New York: Springer-Verlag.

Ohno, S. 1985. Dispensable genes. *Trends in Genet.* 1, 160-164.

Wright, D.A., Richards, C.M., Frost, J.S., Camozzi, A.M. and Kunz, B.J. 1983. *Genetic Mapping in Amphibians*. Vol. 10: Isozymes: current topics in biological and medical research. New York: Alan R. Liss, Inc. 287-311.

The genetic basis of testis determination in man

A. de la Chapelle

Department of Medical Genetics, University of Helsinki, 00290 Helsinki, Finland

The mechanisms by which testes can arise in the apparent absence of a Y chromosome are now becoming clarified. Evidence that Y chromosome-specific DNA occurs in the genomes of XX males has been produced (Guellaen et al. 1984; Vergnaud et al. 1986). It can be reasonably assumed that the testis determining factor (TDF) is amongst those DNA sequences that occur in many XX males. As only a small portion of Yp DNA is necessary to bring about maleness there is reason to believe that TDF alone is responsible for maleness. At any rate present evidence does not suggest that sequences from Yq and/or autosomes are involved in the determination of maleness. A current major challenge for molecular biologists is indeed to further map, characterize, and clone TDF, whose nature is totally unknown. This gene should be of general biological interest in that it may be highly conserved and thus identical or similar between species. In this brief update on current research we shall deal with a number of aspects of XX maleness that still need clarification.

Y DNA OCCURS IN MANY BUT NOT ALL XX MALES

In the study by Vergnaud et al. (1986) the number of XX males tested for the presence of sequences normally occurring on the short arm of the Y chromosome was 19. The results showed that 12 of them had Y DNA while 7 did not. In a study using essentially similar probes, Müller et al. (1986) found Y DNA in 5 out of 7 males with the karyotype 46,XX. The corresponding figures reported by Seboun et al. (1986) were 24 out of 40. Combining these results, Y DNA has so far been found in 41 of the 66 patients tested (62 per cent). Thus it appears clear that while a majority of XX males have DNA originating from the Y chromosome, there is a group of XX males in whom such DNA either does not occur or is not detectable with presently available probes.

THE MECHANISM OF TESTIS DETERMINATION IN XX MALES WHO DO NOT HAVE DNA ENCODED ON Yp

Some or all of these Y DNA negative (Y-) XX males may well have TDF in their genomes. This is obvious when one considers the distribution of Y DNA in those XX males who are DNA positive (Y+). As shown by Vergnaud et al. (1986) the Y DNA sequences can be seen as an overlapping half pyramid indicating the presence of some sequences in all Y+ probands, other sequences in fewer and still other sequences in very few XX males. This pattern indicates that TDF should be closest to the sequence (e.g. p47a) found in the greatest number of probands. Assuming that maleness in at least some Y- probands is due to the presence of TDF, then TDF lies outside of the region spanned by the probes tested for, that is most probably distal of p47a (Fig.1). It follows that some of the patients who score Y- with presently used probes would score Y+ if probes distal of 47a were available. If this hypothesis is correct, XX males will eventually be found who were Y- when tested with presently available probes, but score Y+ with new probes located distal of p47a.

It should also be considered if some XX males arise without having TDF. Other putative mechanisms include mosaicism involving a cell line containing a structurally normal Y chromosome. If such a cell line were extremely circumscribed or hidden, that is, restricted to very few cells, or just occurred in a specific tissue such as testis, then it might escape detection by both cytogenetic and molecular methods of study. The same would hold for a cell line that had occurred in fetal life (accomplishing testis determination) but later had become very circumscribed or perhaps disappeared altogether. The question of mosaicism has been reviewed previously (de la Chapelle 1972; 1981). So far, no compelling evidence indicates that mosaicism is a common etiological factor in maleness without Y. However, occasional "XX" males may indeed be mosaics in analogy with occasional "45,X" males (de la Chapelle et al. 1986). The older literature contains descriptions of at least 4 male patients who were mainly XX but did have a small number of cells that were considered to be 47,XXY (reviewed by de la Chapelle 1972, Table 9). However, these patients were studied before the banding era, so errors of interpretation may have occurred. Authors using the fluorescent Y chromatin in interphase nuclei as a marker have claimed to have found circumscribed mosaicism in a small number of XX males (reviewed by de la Chapelle 1981), but the evidence is not totally convincing (see de la Chapelle et al. 1977 for a discussion of Y chromatin).

Another mechanism leading to maleness in the apparent absence of a Y chromosome is the effect of a putative gene located on the X chromosome or an autosome that turns into a testis determining one as a result of a mutation. For such a gene to exist, familial cases of XX maleness should occur. The number of familial cases reported is in fact remarkably small (reviewed by de la Chapelle 1981). One of the few pedigrees with more than one XX male (de la Chapelle et al. 1977; 1978) is informative in this regard. As shown by molecular studies, all three XX males belonging to this pedigree are Y+, so their maleness is very likely due to the presence of TDF raher than to a gene mutation (Page et al. 1985). How the two second cousins have acquired the same portion of the Y chromosome is not yet clear, however (cf. de la Chapelle 1986 for a discussion).

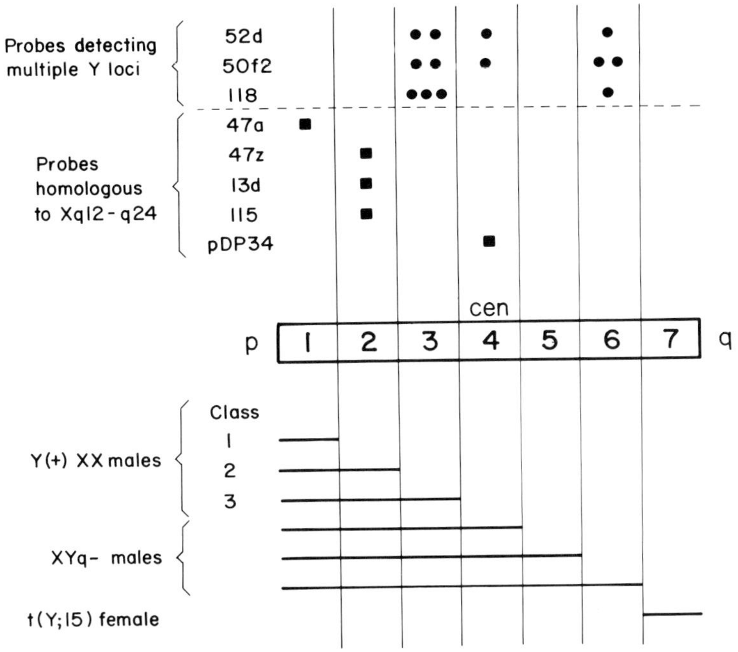

FIGURE 1. Schematic map of the human Y chromosome. From Vergnaud et al. (1986).

ABNORMAL Y-X INTERCHANGE IN THE ETIOLOGY OF XX MALENESS

The mechanism by which Y DNA becomes transferred to occasional XX individuals is in all likelihood an exchange between the Y and X chromosomes in the meiosis of the father. Facts supporting this view are: (1) that XX males have one paternal and one maternal X chromosome (Page and de la Chapelle 1984). (2) The existence of a so-called pseudoautosomal region at the distal ends of the human X and Y chromosomes in which recombination is frequent in male meiosis even under normal circumstances (Simmler et al. 1985; Rouyer et al. 1986) so that (3) if recombination takes place at points proximal of the pseudoautosomal region, TDF might easily get involved. (4) Many XX males do not express the Xg gene of their father suggesting the possibility that Xg is translocated on to the Y chromosome as TDF is translocated on to the X chromosome (reviewed by de la Chapelle 1981). (5) In one XX male proband it was shown that the paternal Xg allele was lost while the postulated homologous Yg gene was present in an XX male child (de la Chapelle et al. 1984). This suggests that at the father's meiosis a portion of Y DNA comprising TDF and Yg was exchanged with a portion of X DNA comprising Xg. (6) In three XX males the Y DNA was present on the terminal portion of the short arm of one X chromosome as shown by in situ hybridization (Andersson et al. 1986). Thus the physical localization of Y DNA in XX males is as predicted if the mechanism is meiotic interchange in the vicinity of the pseudoautosomal region.

ABNORMAL Y-X INTERCHANGE INVOLVES DIFFERENT BREAKPOINTS

As shown in Fig. 1, different classes of XX males can be distinguished based on how much Y DNA their genomes contain. We already hypothesized that TDF is distal of the most distal probes tested for, e.g. 47a. Thus the gene order on Yp is

Ypter - pseudoautosomal - TDF - 47a - 115 - 52d/B - pDP34

just listing a few of those probes that most XX males have been tested for. As different XX males have different sets of Y DNA sequences it is clear that the proximal breakpoint on Yp varies between patients. How about the distal breakpoint? So far no data throwing light on this question have been published. However, such results will soon be forthcoming. The probes emanating from the pseudoautosomal region show multiple polymorphisms (Simmler et al. 1985; Cooke et al. 1985; Rouyer et al. 1986). This will make it possible to determine what part of the distal end of the father's Y chromosome is present in XX males. However, as these pseudoautosomal sequences undergo frequent recombination under normal circumstances in the human male meiosis, it must also be established where the breakpoints occur in the meioses leading to the formation the normal paternal X chromosome of normal females.

Equal and terminal exchange

The simplest model is that both normal and abnormal Y-X interchange is equal and terminal, that is, that only one breakpoint occurs, and that it always occurs between homologous sequences. According to this model, XX males arise when the point of crossing over is so much more proximal than normal that it is proximal of TDF. This model requires that all those RFLP alleles that are distal of TDF on the father's Y chromosome must occur in the XX male son. For normal meioses this model predicts that the more distal paternal Y chromosomal alleles occur more often in the X chromosomes of the daughters than the more proximal alleles. The first data available on normal meioses appear to bear out this hypothesis (Rouyer et al. 1986).

Equal and interstitial exchange

Another model must also be considered. Breakage and reunion may take place in more than one place leading to the meiotic transfer from the Y to the X of an interstitial piece of DNA. According to this model, the paternal X chromosome of an XX male might have its telomeric sequences from the paternal X, then an interstitial portion from the Y comprising TDF, and then again X chromosomal sequences.

No direct evidence for this model has so far been put forward, but data on the inheritance of the blood group Xg provide findings that are compatible with it. As shown by Sanger et al. (1971) and reviewed by de la Chapelle (1981) several XX males are known not to have inherited their father's Xg allele. Assuming that Xg and Yg are homologous sequences and the gene order on the Y is

Ypter - Yg - TDF - Ycen

then the lack of expression of the paternal Xg in several XX males is compatible with the terminal exchange hypothesis. However, other XX males are known who did inherit the paternal Xg (reviewed by de la Chapelle 1981; 1986). If the above gene order is correct, then the gene order on the paternal X chromosome of these XX males might be

Xpter - Xg - TDF - Xcen

which could only have arisen through two crossover events leading to interstitial interchange. While data in support of or against these hypotheses have not yet been published, experiments using 3-generation families or somatic cell hybridization techniques to determine the linkage phase of the pseudoautosomal RFLPs in XX male families are under way. The results will hopefully answer many questions. A reasonable hypothesis is that both terminal and interstitial crossovers occur.

Unequal exchange

Finally, the above two models do not account for unequal crossovers. In the pseudoautosomal region showing very high recombinational activity ("obligatory crossover") it would perhaps not be surprising if occasional mistakes leading to unequal crossovers would occur, and TDF might become involved. Such events would very well explain the variable inheritance of the paternal Xg gene referred to above. Detailed studies of the inheritance of RFLPs emanating from within and proximal of the pseudoautosomal region will answer the question whether or not unequal crossovers occur.

THE MOLECULAR BASIS OF CLINICAL SYMPTOMS IN XX MALES

As more becomes known about the DNA sequences involved in testis determination, it becomes of interest to try to correlate clinical and molecular findings in patients with abnormal sex differentiation. In other words, can anything be learned about the effect of X- and Y-chromosomal genes on the phenotype? A comparison between normal XY males, XX males, and XXY males gives some interesting hints. As females do not have testes while XY and XXY males do, testis determination is indeed due to the Y chromosome. As XX males also have testes but only a tiny portion of the Y chromosome, it is this portion that is responsible for testis determination. More specifically, the putative gene TDF contained in this portion is responsible for testis determination. The dysgenesis of the testes is not due to the lack of other Y chromosomal sequences, because XXY males have equally dysgenetic testes even though they have an entire Y chromosome. Instead, it is the presence of two X chromosomes that leads to the abnormal development of the testes.

Considering body size (height and e.g. tooth size) the following tentative conclusions can be made. As shown by de la Chapelle (1972), height is distributed as follows in order of increasing height

XX women - XX males - XY males - XXY males

The comparison between XY and XXY males indicates that increased dosage of X increases height and that the hormonal disturbances present in XXY males do not lead to diminished height. The comparison between XX males and XXY males suggests that other Y chromosomal genes than TDF increase body size as predicted earlier (Alvesalo and de la Chapelle, 1981). Indeed, the presence of a gene affecting tooth size (and body size) and located in Yq11 has been suggested (Alvesalo and de la Chapelle, 1981). This putative gene has been termed GCY for Growth Control Y chromosome influence (Goodfellow et al. 1985). Finally, the difference in height between XX females and XX males could be due to a number of factors. The marked difference in androgen and other hormonal parameters would appear to be one likely factor, but the effect of genes cannot be dismissed. As the genetic differences between XX females and XX males are quantitatively small (presence of TDF and absence of a minute X chromosomal segment in XX males), such genes, if they exist at all, must be located towards the tips of Xp or Yp.

Further research will determine what other X and Y chromosomal genes affect phenotypic features in man. Such studies regarding genes on the X chromosome are quite difficult because of the inactivation of most of all X chromosomes except one. However, in the case of the Y chromosome a striking feature is already apparent. XX males and XXY males differ in their genetic constitution in that XX males lack most of the Y chromosome. Yet the two clinical conditions are remarkably similar suggesting that very few Y chromosomal sequences (except TDF) have any detectable effect on the phenotype of man.

ACKNOWLEDGEMENTS

This study was supported by grants from the Finska Läkaresällskapet, the Sigrid Jusēlius Foundation and the Academy of Finland. Part of the work was done at the Folkhälsan Institute of Genetics, Finland.

REFERENCES

Alvesalo, L. and A. de la Chapelle 1981.Tooth sizes in two males with deletions of the long arm of the Y chromosome. Ann. Hum. Genet. 54:49-54.

Andersson, M., D.C. Page, and A. de la Chapelle 1986.-Chromosome Y-specific DNA is transferred to the short arm of the X-chromosome in human XX males. Science 233:386-390.

Cooke, H.J., W.R.A. Brown and G.A. Rappold 1985.Hypervariable telomeric sequences from the human sex chromosomes are pseudoautosomal. Nature 317:687-692.

de la Chapelle, A. 1972. Nature and origin of males with XX sex chromosomes. Am. J. Hum. Genet. 24:71-105.

de la Chapelle, A., J. Schröder, J. Murros and G. Tallqvist 1977.Two XX males in one family and additional observations bearing on the etiology of XX males. Clin. Genet. 11:91-106.

de la Chapelle, A., G.C. Koo and S.S. Wachtel 1978.Recessive sex-determining genes in human XX male syndrome. Cell 15:837-842.

de la Chapelle, A. 1981.The etiology of maleness in XX men. Hum. Genet. 58:105-116.

de la Chapelle, A., P.A. Tippett, G. Wetterstrand and D. Page 1984.Genetic evidence of X-Y interchange in a human XX male. Nature 307:170-171.

de la Chapelle, A., D.C. Page, L. Brown, U. Kaski, T. Parvinen and P.A. Tippett 1986.The origin of 45,X males. Am. J. Hum. Genet. 38:330-340.

de la Chapelle, A. 1986.Sex reversal: Genetic and molecular studies on 46,XX and 45,X males. In: Molecular Biology of Homo Sapiens. Cold Spring Harbor Symposium on Quantitative Biology, (in the press).

Goodfellow, P.N., K.E. Davies, and H.H. Ropers 1985.Human Gene Mapping 8. Report of the committee on the genetic constitution of the X and Y chromosomes. Cytogenet. Cell Genet. 40:296-352.

Guellaen, G., M. Casanova, C. Bishop, D. Geldwerth, G. Andre, M. Fellous, and J. Weissenbach 1984.Human XX males with Y single-copy fragments. Nature 307:172-173.

Müller, U., M. Lalande, T. Donlon and S.A. Latt 1986.-Moderately repeated DNA sequences specific for the short arm of the human Y chromosome are present in XX males and reduced in copy number in an XY female. Nucl. Acid. Res. 14:1325-1340.

Page, D.C. and A. de la Chapelle 1984.The parental origin of X chromosomes in XX males determined using restriction fragment length polymorphisms. Am. J. Hum. Genet. 36:565-575.

Page, D.C., A. de la Chapelle, and J. Weissenbach 1985.-Chromosome Y-specific DNA in related human XX males. Nature 315:224-226.

Rouyer, F., M.-C. Simmler, C. Johnsson, G. Vergnaud, H.J. Cooke, and J. Weissenbach 1986.A gradient of sex linkage in the pseudoautosomal region of the human sex chromosomes. Nature 319:291-295.

Sanger, R., P. Tippett, and J. Gavin 1971.The X-linked blood group system Xg: tests on unrelated people and families of Northern European ancestry. J. Med. Genet. 8:427-433.

Seboun, E., P. Leroy, M. Casanova, E. Magenis, C. Boucekkine, C. Disteche, C. Bishop and M. Fellous 1986.A molecular approach to the study of the human Y chromosome and anomalies of sex determination in man. In: Molecular Biology of Homo Sapiens. Cold Spring Harbor Symposium on Quantative Biology, (in the press).

Simmler, M.-C., F. Rouyer, G. Vergnaud, M. Nyström-Lahti, K.Y. Ngo, A. de la Chapelle and J. Weissenbach 1985. Pseudoautosomal DNA sequences in the pairing region of the human sex chromosomes. Nature 317:692-697.

Vergnaud, G., D. C. Page, M.-C. Simmler, L. Brown, F. Rouyer, B. Noel, D. Botstein, A. de la Chapelle and J. Weissenbach 1986.A deletion map of the human Y chromosome based on DNA hybridization. Am. J. Hum. Genet. 38:109-124.

A molecular analysis of the human Y chromosome

J. Weissenbach

INSERM U.163, Institut Pasteur, 75724 Paris Cédex 15, France

The differentiation of individuals of the same species into females and males is achieved by a sophisticated developmental process. In mammals this process is under genetic control (Eicher, 1986) and it is supposed that one (or more) putative loci residing on the Y chromosome code for a factor(s) (TDF) triggering testis differentiation.

Identifying such loci on the Y chromosome appears difficult since mutations of genes involved in sex determination impair gonadal differentiation and may therefore not be transmittable. In man, in contrast to all other chromosomes, only a single gene has been mapped to the human Y chromosome (Goodfellow et al., 1983). Strikingly this locus maps also to the X chromosome, illustrating that the haploid state of most of the Y chromosome has hampered considerably the establishment of an informative genetic map.

In recent years, isolation of Y-specific DNA sequences (Bishop et al., 1983) has provided a new tool for the structural analysis of the Y chromosome and several related anomalies e.g. XX maleness (Guellaen et al.). This has enabled investigators to propose a deletion map of the human Y chromosome (Vergnaud et al.). More recently, crossing over between the human X and Y chromosomes has been demonstrated using other Y-derived probes (Cooke et al., 1985; Simmler et al., 1985; Rouyer et al., 1986a). These two major aspects of the recent developments in the molecular biology of the human Y chromosome are reviewed below.

ISOLATION AND CHARACTERIZATION OF Y-SPECIFIC SEQUENCES

Y-located DNA probes can detect DNA fragments which either appear exclusively in male DNA and have no homology with other parts of the genome (Y-limited DNA fragments) or are specific for the Y chromosome though homologous to a certain extent with sequences located on other chromosomes (Y-specific fragments). Whereas Y-specific repetitive DNA fragments have been isolated readely some time ago (Cooke et al., 1976; Kunkel et al., 1976), the first probes for Y-specific single copy DNA were obtained as a serendipic outcome of testing probes from various origins (Page et al., 1982; Daiger et al., 1982). Numerous Y-specific and Y-limited probes have been obtained since from cosmid or lambda phage libraries derived from human/rodent cell hybrids in which the human Y chromosome was the only detectable human genomic material (Bishop et al., 1983, 1984; Wolfe et al., 1984; Burk et al., 1985). The clones containing human DNA were detected on the basis that they hybridize specifically to human repetitive DNA and not to murine repetitive DNA (Gusella et al., 1980). Y-specific DNA probes were then isolated after cloning subfragments devoid of repetitive DNA. A more recent approach used clones from phage libraries of flow sorted chromosomes (Affara et al. 1986a; Muller et al., 1986; Page et al., 1986).

Several criteria are used to characterize the fragments detected by these probes: (i) presence of one or more male-specific fragments, (ii) chromosomal mapping of cognate and cross-hybridizing bands using appropriate somatic cell hybrids, (iii) dosage effect on the hybridizing intensity of male specific or cross-hybridizing fragments in DNA digests from cell lines carrying multiple sex chromosomes, (iv) in situ hybridization to chromosome metaphase spreads. These characteristics provide some information in regard to the chromosomal relationships of Y-located DNA fragments. It has notably been observed that 20% to 30% of non tandemy repeated Y-located DNA material is highly homologous to sequences located between Xq12 and Xq24 (Geldwerth et al., 1985). These sequences were most probably translocated from the X chromosome long arm to the Y chromosome short arm during recent human evolution as has been shown for some of them (Page et al. 1984; Koenig et al., 1984). It has also been observed that a few Y DNA single copy fragments share homologies with autosomal DNA sequences (Bishop et al., 1984; Affara et al., 1986a) or are pseudogenes (Daiger et al., 1982; Heilig et al., 1984). In addition Y-limited DNA fragments occur often in several interspersed copies mapping to either one or both arms of the chromosome (Vergnaud et al., 1986).

Y-SPECIFIC DNA IN SEX DETERMINATION ANOMALIES

The newly characterized Y-specific DNA probes can be used to investigate anomalies of sex determination like human XX maleness or pure gonadal dysgenesis (XY females). Human XX females are sterile individuals with male gonads and a 46,XX karyotype (de la Chapelle, 1981). The presence of Y-specific DNA was first shown in three out four males (Guellaen et al., 1984). Such patients with Y-specific DNA are noted as Y(+) XX males. This analysis also suggested that these three Y(+) cases had inherited a portion of the Y chromosome which was variable in size. More extensive studies have confirmed these earlier observations and are consistent with the fact that each Y(+) XX male carries a single contiguous segment of the Y chromosome (Vergnaud et al., 1986; Muller et al., 1986; Affara et al., 1986b; Page et al., 1986). These patients carry Y-specific fragments originating from Yp. With a few exceptions (Affara et al., 1986b; Vergnaud et al., unpublished), the Y chromosomal segments of the different cases can be arranged in an inclusive overlapping series (nested series), in which, among two different segments, the larger one completely overlaps the smaller one (Vergnaud et al., 1986; Muller et al., 1986; Affara et al., 1986b; Page et al., 1986). Taken together these results strongly suggest an oriented transfer of a portion of the Y chromosome short arm and are consistent with a transfer including the Yp telomere (Vergnaud et al., 1986; Page et al., 1986).

In situ hybridization experiments with Y-specific DNA probes have confirmed that the Y-specific fragments present in Y(+) XX males map to Yp and have shown that this portion is transferred to the tip of Xp in these patients (Casanova et al., 1985; Magenis et al., 1985; Buckle et al., 1985a; Muller et al., 1986; Anderson et al., 1986). It has been confirmed that in the Y(+) cases of XX maleness analysed to date, the telomere originates from the paternal Y chromosome and is associated to a loss of the paternal Xp telomere (Page et al., 1986; Rouyer et al., 1986; Petit et al., in preparation). These results support Ferguson-Smith's hypothesis (1964) proposing that XX males result from an abnormal interchange between the X and Y chromosomes. Thus human Y(+) XX maleness shows some similarities with the mouse Sxr mutation (Singh and Jones, 1982; Evans et al., 1982) in that both seem to involve an aberrant genetic exchange between sex chromosomes at paternal meiosis.

A deletion map (FIGURE 1a) of the human Y chromosome has been constructed assuming that the Y(+) XX males arose from an abnormal X/Y interchange involving the terminal part of Yp. The different Y(+) XX males define several classes carrying each a Y chromosomal portion of a different size. The deletion map was completed by addition of surnumerary blocs deduced from the hybridization pattern of Y-specific probes to other structural anomalies of the human Y chromosome (Yq deletions, duplications, translocations etc...). The long arm telomere and the centromere were assigned to defined intervals of a first map comprising seven contiguous deletion blocs. The putative testis determining factor belongs to the most distal Y-specific interval on the short arm common to all classes of Y(+) XX males. This interval map has been refined recently by Page et al. (1986) and comprises 12 intervals to date. The pseudoautosomal region (see below) which corresponds to the terminal part of Yp is not included in the deletion map, but is adjacent to the most distal interval of the sex-specific part of the chromosome (FIGURE 1b).

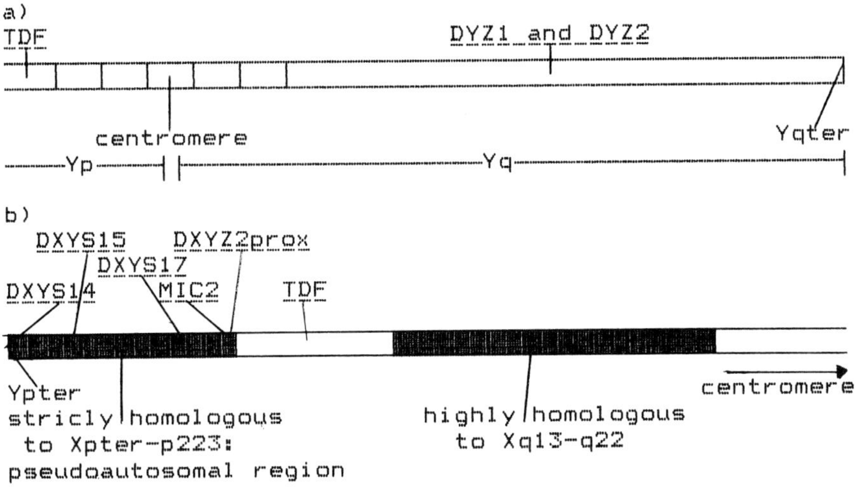

FIGURE 1a) A deletion map of the human Y chromosome with seven intervals. 1b) Structural regions of Yp. Sizes of the different intervals and regions do not correspond to physical distances.

Other anomalies like XY pure gonadal disgenesis have been studied more recently. Although a smaller number of cases was analysed, the results indicate that the Y chromosomal short arm of several such cases has been partially deleted and that the Y-specific fragments involved in these deletions correspond to fragments present in Y(+) XX males (Casanova et al., 1985; Muller et al., 1986; Page et al. 1986). Thus XX maleness and XY femaleness may both originate from abnormal interchanges at male meiosis and could represent reciprocal abnormal events. This possibility appears even more likely with the recent demonstration of a normal crossing over between the human X and Y chromosomes.

CROSSING OVER BETWEEN THE HUMAN SEX CHROMOSOMES

Crossing over between the mammalian sex chromosomes has been proposed more than fifty years ago by Koller and Darlington (1934). Although several theoretical arguments and cytological observations supporting this view have been reported since (Solari, 1980; Polani, 1982, Burgoyne, 1982; Holm and Rasmussen, 1983; Chandley et al., 1985), experimental evidence for crossing over has only been produced recently. It has been shown that in mouse, the enzyme steroid sulfatase is encoded by a gene mapping to the X chromosome. However this enzymatic activity does not show sex linkage at male meiosis and is therefore supposed to segregate via an obligatory recombination between the X and Y chromosomes (Keitges et al., 1985). This observation is consistent with earlier results showing that the Sxr mutation residing on an abnormal Y chromosome is transferred to the paternal X chromosome in 50% of the XX caryotyped progeny (Singh and Jones, 1982; Evans et al., 1982).

In man, numerous DNA loci mapping to the human Y-chromosome and sharing homology with X-located DNA sequences have been isolated (Goodfellow et al., 1985). But in all earlier instances reported either the DNA locus from the X or from the Y or from both sex chromosomes did not map to the apparent pairing region on the short arms. Since 1985 a few DNA probes have shown sequences homologies located in the pairing region of both the X and Y chromosomes (Cooke et al., 1985; Simmler et al., 1985; Rouyer et al., 1985, 1986; Buckle et al., 1985b; Page et al., 1986; Affara et al., 1986a). These sequences detected also RFLPs which could be used in family studies to test for sex linkage. Most of these loci could recombine with sexual phenotype (Cooke et al., 1985; Simmler et al., 1985; Rouyer et al., 1985, 1986; Goodfellow et al., 1986; Page et al., 1986), giving experimental support to the concept of pseudoautosomal loci proposed by Burgoyne (1982).

These results have now been extended to five distinct pseudoautosomal loci in a quantitative family study on more than hundred male and female meioses (Rouyer et al., 1986). This linkage analysis first shows that different pseudoautosomal loci recombine at different frequencies (TABLE 1), according to a gradient of sex linkage in the human pseudoautosomal region. The telomeric locus DXYS14 recombines with sex at a frequency of 50% and is thus not sex-linked, whereas all other pseudoautosomal loci analysed exhibit all partial sex linkage. This sex linkage gradient can be represented on a linear map, were the five loci can be ordered with respect to their recombination distances with the X and Y sex-specific chromosomal blocks.

TABLE 1. Recombination between pseudoautosomal loci and sexual phenotype.

loci	meioses	recombinations	Recombination frequency
DXYS14	114	57	50%
DXYS15	82	29	35.5%
DXYS17	86	10	11.5%
MIC2	37	1	2.5%
DXYZ2prox	46	1	2.2%

Examining cotransmission of several loci shows that in all instances studied, the less frequently recombined locus never segregates independently from loci recombining more frequently (Rouyer et al., 1985, 1986). This indicates that X/Y interchange of pseudoautosomal loci results from a crossing over and not from other genetic exchange events, such as gene conversion, through which the different loci would segregate at random. These results also strongly suggest that a single event of crossing over occurs between the X and Y chromosomes since no examples of double recombination have been found in 114 analysed meioses. As different pseudoautosomal loci recombine at different frequencies, it is concluded that the crossover point is not uniquely localized.

These features are confirmed by measurements of recombination frequencies between different pseudoautosomal loci. In all instances it appears that the sum of the recombination frequencies of two or more adjacent intervals equals the frequency measured directly between the extreme loci of the superimposing interval (TABLE 2). Thus the classical underestimation of larger genetic distances does not occur and the rule of additivity of recombination frequencies, which usually applies to the 0-15% range, is valid for the entire pseudoautosomal region which stretches on 50 recombination units. This property characterizes the human pseudoautosomal region and represents strong additional evidence in favor of a single crossing over between the human sex chromosomes per male meiosis.

TABLE 2. Recombination frequencies between pseudoautosomal loci.

Intervals	male meiosis	female meiosis
DXYS14-DXYS15	13.5	0
DXYS14-DXYS17	37	2.5
DXYS14-MIC2	48.5	2
DXYS14-DXYZ2prox	54.5	5.3
DXYS15-DXYS17	24	3.2
DXYS15-MIC2	35	3.2
DXYS15-DXYZ2prox	37	0
DXYS17-MIC2	11	0
DXYS17-DXYZ2prox	7.4	5.3
MIC2-DXYZ2prox	0	2.8

In addition a dramatic decrease of recombination frequency in the pseudoautosomal region at female meiosis as compared to male meiosis can be observed (TABLE 2). A total of four events has been detected out of hundred female meioses, whereas 50% of male gametes are recombined in the pseudoautosomal region. This contrasts with a significant increase reported for autosomal recombination frequency in female versus male meiosis (Renwick, 1968). This striking difference is consistent with the view that a single crossover between the mammalian X and Y chromosomes is obligatory at male meiosis. If, as proposed by Koller and Darlington (1934), one chiasma at least takes place in each bivalent to ensure proper segregation during the first meiotic division, then the 10 to 20 fold increase in male recombination frequency is a direct consequence of a chiasma having to be formed in an as limited chromosomal segment as the pseudoautosomal region.

CONCLUSION

Isolation and use of Y-specific and pseudoautosomal DNA probes has allowed considerable progress in our present knowledge of the human Y chromosome. On the one hand, an accurate regional mapping of the testis determinant to the distal part of Yp has been achieved. On the other hand, occurrence at the tip of the short arms of the X and Y chromosomes of a pseudoautosomal region, in which loci can undergo recombination between the two sex chromosomes at male meiosis has been definitely demonstrated.

Numerous questions remain so far unsolved. In spite of the intensive efforts of several groups, no human Y-limited transcription product has been evidenced yet. A refined mapping of TDF is also awaited. The evolution of mammalian sex chromosomes represents an other aspect deserving further insight. This question can now be addressed with the numerous Y-specific and pseudoautosomal DNA probes available. But carrying out a physical map using procedures of pulsed field gel electrophoresis should be the next major endeavour for the human Y chromosome.

REFERENCES

Affara, N.A., L. Florentin, N. Morrison, K. Kwok, M. Mitchell, A. Cook, D. Jamieson, L. Glasgow, L. Meredith, E. Boyd and M.A. Ferguson-Smith 1986a. Regional assignment of Y-linked DNA probes by deletion mapping and their homology with X-chromosome and autosomal sequences. Nucleic Acids Res. 14, 5353-73.

Affara, N.A., M.A. Ferguson-Smith, J. Tolmie, K. Kwok, M. Mitchell, D. Jamieson, A. Cooke and L. Florentin 1986b. Variable transfer of Y-specific sequences in XX males. Nucleic Acids Res. 14, 5375-87.

Anderson M., D.C. Page and A. de la Chapelle 1986. Chromosome Y-specific DNA is transferred to the short arm of X chromosome in human XX males. Sience 233, 786-8.

Bishop, C.E., G. Guellaen, D. Geldwerth, R. Voss, M. Fellous and J. Weissenbach 1983. Single-copy DNA sequences specific for the human Y chromosome. Nature 303, 831-2.

Bishop, C., G. Guellaen, D. Geldwerth, M. Fellous and J. Weissenbach 1984. Extensive sequence homologies between Y and other human chromosomes. J. Mol. Biol. 173, 403-17.

Buckle, V., Y. Boyd, I.W. Craig, N. Fraser, P.N. Goodfellow and J. Wolfe 1985a. Localization of Y chromosomal sequences in normal and XX males. Cytogenet. Cell Genet. 40, 593.

Buckle, V., C. Mondello, S. Darling, I.W. Craig and P.N. Goodfellow 1985b. Homologous expressed genes in the human sex chromosomes pairing region. Nature 317, 739-41.

Burgoyne, P.S. 1982. Genetic homology and crossing over in the X and Y chromosomes of mammals. Hum. Genet. 61, 85-90.

Burk, R.D., P. Ma and K.E. Smith 1985. Characterization and evolution of single copy sequence from the human Y chromosome. Mol. Cell Biol. 5, 576-81.
Casanova, M.E., E. Seboun, P. Leroy, C. Junien, I. Henry, E. Boucekkine, E. Magenis and M. Fellous 1985. Yp encoded sequences in the short arm of paternal X in human XX males and XX males with true hermaphroditism. Cytogenet. Cell. Genet. 40, 600.
Chandley, A.C., P. Goetz, T.B. Hargreave, A.M. Joseph and R.M. Speed 1984. On the nature and extent of XY pairing at meiotic prophase in man. Cytogenet. Cell Genet. 38, 241-7.
Cooke, H.J., 1976. Repeated sequences specific for human males. Nature 262, 182-6.
Cooke, H.J., W.A.R. Brown and G. Rappold 1985. Hypervariable telomeric sequences from the human sex chromosomes are pseudoautosomal. Nature 317, 688-92.
de la Chapelle, A. 1981. The etiology of maleness in XX men. Hum. Genet. 58, 105-16.
Daiger, S.P., R.S. Wildin and T.S. Su 1982. Sequences on the human Y chromosome homologous to the autosomal gene for argininosuccinate synthetase. Nature 298, 682-4.
Eicher, E.M. and L.L. Washburn 1986. Genetic control of primary sex determination in mouse. Ann. Rev. Genet. in press.
Evans, E.P., M.D. Burtenshaw and B.M. Cattanach 1982. Meiotic crossing over between the X and Y chromosomes of male mice carrying the sex reversing (Sxr) factor. Nature 300, 443-5.
Ferguson-Smith M.A. 1966. X-Y chromosomal interchange in the aetiology of true hermaphroditism and of XX Klinefelter's syndrome. Lancet ii, 475-6.
Goodfellow, P.N., C. Banting, D. Sheer, H.H Ropers, M.A. Caine, M.A. Ferguson-Smith, S. Povey and R. Voss 1983. Genetic evidence that a Y-linked gene in man is homologous to a gene on the X chromosome. Nature 302, 346-9.
Goodfellow, P.N., K.E. Davies and H.H. Ropers 1985. Report of the committee on the genetic constitution of the human X and Y chromosomes. Cytogenet. Cell Genet. 40, 296-352.
Goodfellow, P.J., S.M. Darling, T.N. Thomas and P.N. Goodfellow 1986. A pseudoautosomal gene in man. Sience in press.
Guellaen, G., M. Casanova, C. Bishop, D. Geldwerth, G. André, M. Fellous and J. Weissenbach 1984. Human XX males with Y single-copy DNA fragments. Nature 307, 172-3.
Gusella, J.F., C. Keys, A. Varsanyi-Breiner, F.T. Kao, C. Jones, T.T. Puck and D. Housman 1980. Isolation and characterization of DNA segments from specific chromosomes. Proc. Natl. Acad. Sci. USA 77, 2829-33.
Heilig, R., A. Hanauer, K.H. Grzeschik, M.C. Hors-Cayla and J.L. Mandel 1984. Actin-like sequences are present on the X and Y chromosomes. EMBO J. 3, 1803-7.
Holm, P.B. and S.W. Rasmussen 1983. Human meiosis VI. Crossing over in human spermatocytes. Carlsberg Res. Comm. 48, 385-413.
Keitges, E., M. Rivest, M. Siniscalco and S.M. Gartler 1985. X-linkage of steroid sulphatase in the mouse is evidence for a functional Y-linked allele. Nature 315, 226-7.
Koenig, M., J.P. Moisan, R. Heilig and J.L. Mandel 1985. Homologies between the X and Y chromosomes detected by DNA probes: localization and evolution. Nucleic Acids Res. 13, 5485-501.
Koller, P.C. and C.D. Darlington 1934. The genetical and mechanical properties of the sex chromosomes 1. *Rattus norvegicus*. J Genet. 29, 159-73.

Kunkel, L.M., K.D. Smith and S.H. Boyer 1976. Human Y chromosome specific reiterated DNA. Science 191, 1189-90.
Muller, U., M. Lalande, T. Donlon and S.A. Latt 1986. Moderately repeated DNA sequences specific for the short arm of the human Y chromosome are present in XX males and reduced in copy number in an XY female. Nucleic Acids Res. 14, 1325-40.
Magenis, R.E., R. Sheehy, S. Olson, M.G. Brown, M. Casanova and M. Fellous 1985. Genes for maleness (TDF) localized to distal one-half of Y chromosome short arm : evidence from in situ hybridization of a Y-derived single copy DNA probe. Cytogenet. Cell Genet. 40, 686.
Page, D., B. de Martinville, D. Barker, A. Wyman, R. White, U. Francke and D. Botstein 1982. Single copy sequence hybridizes to polymorphic and homologous loci on human X and Y chromosomes. Proc. Natl. Acad. Sci. USA 74, 5352-6.
Page, D.C., M.E. Harper, J. Love and D. Botstein 1984. Occurence of a transposition from the X chromosome long arm to the Y chromosome short arm during human evolution. Nature 311, 119-23.
Page, D.C., L. Brown, K. Bieker, J. Pollack. 1986. Sex-reversal: deletion mapping the male-determining function of the Y chromosome. Cold Spring Harbor Symp. Quant. Biol. 51, in press.
Polani, P.E. 1982. Pairing of X and Y chromosomes, non inactivation of X-linked genes, and the maleness factor. Hum. Genet. 60, 207-11.
Renwick, J.H. 1968. Ratios of female to male recombination in man. Bull. Europ. Soc. Hum. Genet. 2, 7-8.
Rouyer, F., M.C. Simmler, C. Johnsson, G. Vergnaud, H. Cooke and J. Weissenbach 1986. A gradient of sex linkage in the pseudoautosomal region of the human sex chromosomes. Nature 319, 291-5.
Rouyer, F., M.C. Simmler, G. Vergnaud, C. Johnsson, C. Petit, J. Levilliers and J. Weissenbach 1986. The pseudoautosomal region of the human sex chromosomes. Cold Spring Harbor Symp. Quant. Biol. 51, in press.
Simmler, M.C., F. Rouyer, G. Vergnaud, M. Nyström-Lahti, K.Y. Ngo, A; de la Chapelle and J. Weissenbach 1985. Pseudoautosomal DNA sequences in the pairing region of the human sex chromosomes. Nature 317, 692-7.
Singh, L. and K.W. Jones 1982. Sex reversal in the mouse (Mus musculus) is caused by a recurrent nonreciprocal crossover involving the X and the aberrant Y chromosome. Cell 28, 205-16
Solari, A.J. 1980. Synaptonemal complexes and associated structures in microspread human spermatocytes. Chromosoma 81, 315-37.
Vergnaud, G., D.C. Page, M.C. Simmler, L. Brown, F. Rouyer, B. Noel, D. Botstein, A. de la Chapelle and J. Weissenbach 1986. A deletion map of the human Y chromosome based on DNA hybridization. Am. J. Hum. Genet. 38, 109-24.
Wolfe, J., R.P. Erickson, P.W.J., Rigby and P.N. Goodfellow 1984. Cosmid clones derived from both euchromatic and heterochromatic regions of the human Y chromosome. EMBO J. 3, 1997-2003.

An evolutionarily conserved early replicating segment on the sex chromosomes of man and the great apes

W. Schempp and B. Weber

Institut für Humangenetik, Universität Freiburg i. Br., Federal Republic of Germany

Regarding the evolution of sex chromosomes, the process of sex chromosome differentiation may be described as a process starting from originally homomorphic and proceeding to highly differentiated heteromorphic sex chromosomes.

An early step of sex chromosome differentiation appears to manifest itself by an asynchrony of DNA-replication: In the amphibian species Rana esculenta the evolution of sex chromosomes must be on a very early level, since no morphologic difference can be detected between homologous chromosomes, even when structural banding techniques are used. A heteromorphism in one chromosome pair,however, can be detected applying the BrdU replication technique (Schempp and Schmid, 1981). While in female animals both chromosomes No. 4 replicate synchronously, male animals exhibit an extremely late replicating interstitial segment in the long arm of one of the chromosomes No. 4 which therefore, must be the Y chromosome. This replication heteromorphism between the X and Y of Rana esculenta can be defined as a very early stage of differentiation of homomorphic chromosomes prior to becoming heteromorphic XX/XY chromosomes.

According to the hypothesis of Ohno (1967), the differentiation of X and Y chromosomes in mammals is assumed to have been initiated by a mechanism which prevents or curtails free crossing-over between these chromosomes. This isolation mechanism would provide

the basis for the differential accumulation of sex determining genes on the originally homomorphic chromosomes. Furthermore, it would permit a morphologic differentiation between these chromosomes, as apparently was the case by loss of genetically inert material in the Y chromosome.
The possibility that homologies exist between the mammalian X and Y chromosomes and that these regions on the X chromosome might escape inactivation was already considered by Lyon (1962, 1970) shortly after advancing her original X-inactivation hypothesis.

That there are in fact such homologies between parts of the X and Y chromosome is now supported by cytologic findings as well as by molecular data on the DNA level.
Since on the cytological level synchronous replication activity in autosomal homologues indicates functional homology, we made use of the BrdU replication technique and studied high resolution replication patterns of prometaphasic human X and Y chromosomes, paying special attention to the distal short arms of the sex chromosomes.
In male cells we were able to detect synchronous early replication on the distal short arms of both prometaphasic X and Y chromosomes (Müller and Schempp, 1982). In another investigation (Schempp and Meer, 1983) we studied the early replication patterns of the sex chromosomes in normal female cells and in cells of Klinefelter probands. The replication behaviour of the distal short arm of the inactivated human X chromosome was studied in an attempt to test the hypothesis that this part of Xp escapes late replication, i.e. inactivation. The cells of Klinefelter males make it possible to study the early and the late X chromosomes as well as the Y chromosome in a single complement.

In summary these studies have revealed a distinct early replicating segment on both distal Xp and Yp, corresponding to the pairing segment in which a synaptonemal complex is formed in male meiosis. Furthermore, this early replicating segment on Xp appears to escape inactivation, supporting the suggestion that the segments on distal Xp and Yp are a remnant left unchanged during the differentiation into heteromorphic sex chromosomes of originally homomorphic autosomes.

The question arises if a similar replication behaviour can be demonstrated for the sex chromosomes of the great apes. Therefore we have performed a comparative replication study on prometaphasic chromosomes paying special attention to the sex chromosomes of man, chimpanzee, pygmy chimpanzee, gorilla and orangutan (Weber et al., in press).

The band by band analysis of the early X chromosomes of man and the great apes clearly demonstrates the high conservation of both the pattern and the time course of replication.
The heterochromatic telomeric "caps" on the gorilla X as well as on chimpanzee Xq can be neglected here because of their nonfunctional character.
On the inactivated X chromosomes of the higher primates a single band located in Xp22.3 each replicates early in the S-phase and therefore should escape inactivation.
According to our cytogenetic data on the Y chromosome of man and the great apes it is evident that in both the euchromatic and heterochromatic parts, structural alterations have resulted in an extreme divergence. It is assumed, therefore that in contrast to the X chromosomes, the Y chromosomes have undergone a rapid evolution within the higher primates.
The earliest replicating segment is situated on the distal Yp of man and orangutan, whereas it is located on the distal long arm of the chimpanzee and pygmy chimpanzee Y chromosomes, respectively.
In the gorilla this earliest replicating segment is located proximally to an additional heterochromatic telomeric cap on Yp.
Similar as in man, in the great apes only the earliest replicating segment each on the Y chromosomes and the synchronously early replicating segment on the early and late X chromosomes may be homologized. Consequently these segments could correspond to the pseudoautosomal segment which was shown to exist in man (Simmler et al., 1985; Cooke et al., 1985). The heterochromatic caps on the short arms of the gorilla sex chromosomes do not interfere with this view. Furthermore, these earliest replicating segments on human Xp and Yp are included in the pairing region in male meiosis.

Therefore we predict that pairing in the chimpanzee and pygmy chimpanzee will take place between Xp and the Y chromosome long arm, while in the other species it will be between Xp and Yp, similar as in man.

Recently human Y chromosome derived sequences which by in situ hybridization were located in the pairing region of human X and Y chromosomes have been shown to undergo recombination and they were therefore described as "pseudoautosomal" (Simmler et al., 1985; Cooke et al., 1985).

In situ hybridization of the human derived pseudoautosomal probe 113F that detects the locus DXYZ2 (Simmler et al., 1985), most clearly demonstrates that this pseudoautosomal sequence was conserved and is located also in the synchronously early replication segments of the great apes sex chromosomes. Therefore these segments may be homologous to the pseudoautosomal segment existing in man.

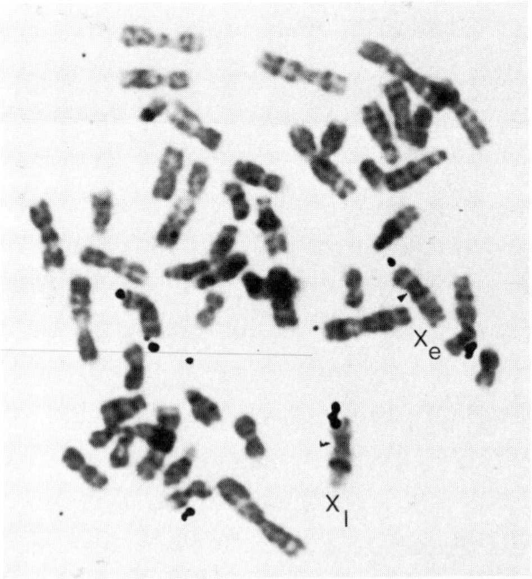

Fig. 1: Metaphase spread of the female chimpanzee with early replication patterns after in situ-hybridization with the pseudoautosomal probe 113F. Note the grains on the distal short arm of both the early (X_e) and the late X (X_l), respectively.

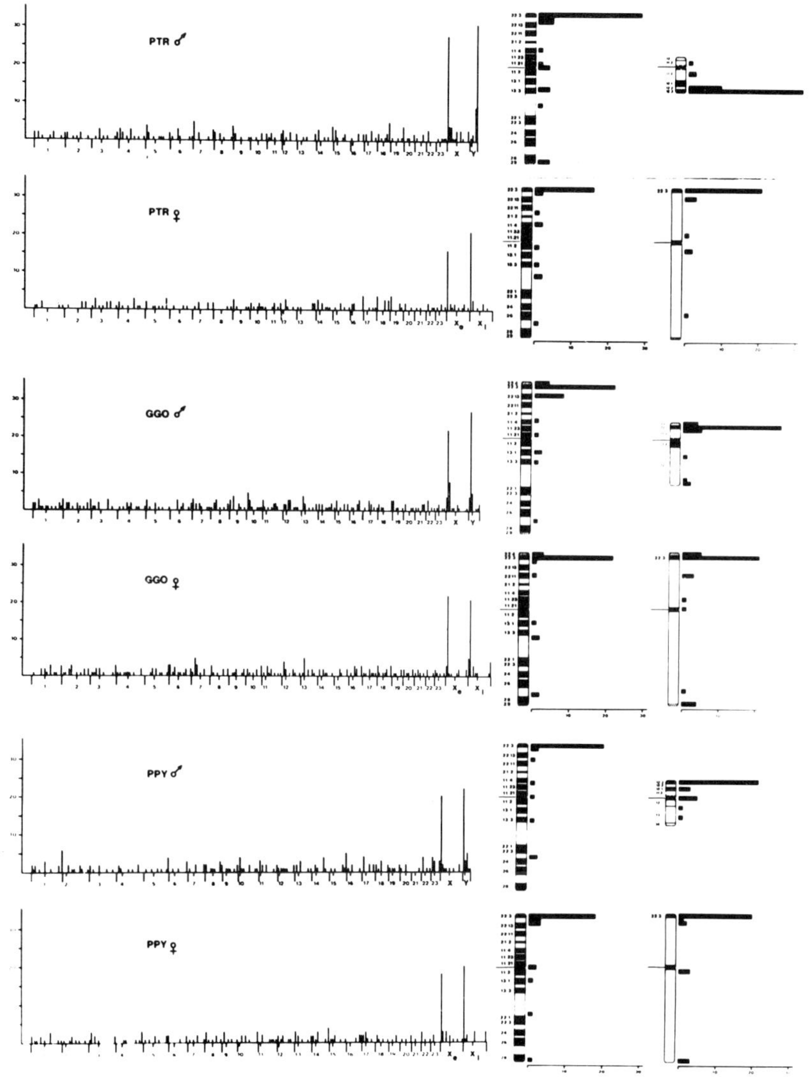

Fig. 2. In situ-hybridization with the pseudoautosomal probe 113F. Histogrammatic representation of the grain distribution on the chromosome complements of chimpanzee (PTR), gorilla (GGO) and orangutan (PPY)(left). A significant grain accumulation can be shown on the sex chromosomes. Detailed grain distribution on the male and female sex chromosomes of chimpanzee, gorilla and orangutan, respectively (right).

For the in situ experiments male and female animals of the chimpanzee (PTR), gorilla (GGO) and orangutan (PPY) were available, respectively.
The analyses of the in situ experiments with the pseudoautosomal probe 113F are summarized in Fig. 2.
According to our most recent results it is quite obvious, that even in the lower primate Macaca mulata this pseudoautosomal probe 113F hybridizes specifically to the sex chromosomes. So far we have analyzed cells from female animals only. Interestingly grain accumulation is clearly visible in the most distal early replicating segment of both X chromosomes in female cells.
As the earliest replicating segment of the rhesus monkey Y chromosome is located on the distal long arm (unpublished results), we will predict that hybridization of the pseudoautosomal probes occurs on distal Yq.

This conservation of pseudoautosomal sequences supports the suggestion that there are indeed homologous segments on the X and Y chromosomes representing a remnant left unchanged during the differentiation into heteromorphic sex chromosomes of originally homomorphic autosomes.
Little is known, however, about the genetic information, i.e. the function of these pseudoautosomal segments on primates sex chromosomes.
It may be speculated that these segments on the sex chromosomes have a role at the formation of a synaptonemal complex at early pachytene of male meiosis.

ACKNOWLEDGEMENT

This work was supported by grants of the Deutsche Forschungsgemeinschaft.

REFERENCES

Cooke, H. J., Brown, W. R. A., Rappold, G. A. 1985.Hypervariable telomeric sequences from the human sex chromosomes are pseudo-autosomal. Nature 317, 687-692.

Lyon, M. F. 1962.Sex chromatin and gene action in the mammalian X chromosome. Am. J. Hum. Genet. 14, 135-148.

Lyon, M. F. 1970.Genetic activity of sex chromosomes in somatic cells of mammals. Philos. R. Soc. Lond. (Biol.) 259, 41-52.

Müller, U. and W. Schempp 1982.Homologous early replication patterns of the distal short arms of prometaphasic X and Y chromosomes. Hum. Genet. 60, 274-275.

Ohno, S. 1967.Chromosomes and sex linked genes. Berlin, Heidelberg, New York: Springer

Schempp, W. and M. Schmid 1981. Chromosome banding in amphibia. VI. BrdU-replication patterns in Anura and demonstration of XX/XY sex chromosomes in Rana esculata. Chromosoma 83, 697-710.

Schempp, W. and B. Meer 1983.Cytologic evidence for three human X-chromosomal segments escaping inactivation. Hum. Genet. 63, 171-174.

Simmler, M. C., Rouyer, F., Vergnaud, G., Myström-Lahti, M., Ngo, K. Y., de la Chapelle, A., Weissenbach, J. 1985. Pseudoautosomal DNA sequences in the pairing region of the human sex chromosomes. Nature 317, 692-697.

Weber, B., Schempp, W., Wiesner, H. An evolutionaryly conserved early replicating segment on the sex chromosomes of man and the great apes. Cytogenet.Cell Genet. in press

MEIOSIS AND ANEUPLOIDY

Meiotic behavior of sex chromosomes: what *is* normal?

T. Ashley

Department of Zoology, University of Tennessee, Knoxville, Tennessee 37996, USA

Molecular techniques of DNA cloning and probing have recently resulted in a deluge of information about DNA sequences on the X and Y chromosomes. It is essential that this information be intergrated into the genetic and cytological information. In this paper known facts will be identified and questions raised that will focus on the areas of disparity between the cytological and molecular data and hopefully provide a direction for future research. Apparent conflicts between these two fields of study must be reconciled if we are to emerge with a clear and coherent picture of the meiotic behavior of the sex chromosomes of mammals.

Points at issue include synapsis (homologous and nonhomologous), crossing over, chiasmata formation, and the mechanism of disjunction. The stage was set 50 years ago by Koller and Darlington (1934) who observed a region of synapsis between the X and Y in the Norwegian rat and an end association of these two chromosomes at metaphase I. They concluded that there was a region of homology between the two and that a crossover occurred in this region which assured normal disjunction. They later concluded XY behavior in all mammals followed this pattern. Although challenging various Darlington theories has been a favorite pastime of cytogeneticists for many years, this particular idea seems to have become entrenched and accepted without the usual scrutiny. I invite you to join me in a close examination of these issues.

Chromosomes today, volume 9 Allen & Unwin 1987

THERE IS EXTENSIVE SYNAPSIS BETWEEN THE X AND Y IN MOUSE, MAN, AND MANY (BUT NOT ALL) OTHER MAMMALS.

Extensive synapsis between the X and Y has been reported for mouse (Tres, 1977), man (Solari, 1980; Chandley, et al., 1984) and many other mammals. On the other hand, there is no reported synapsis in any marsupial with a normal XY sex determining mechanism (Sharp, 1982), nor in several eutherian mammals that have been examined including the sand rat (Solari and Ashley, 1977), and the pygmy mouse (Pathak, et al., 1980).

IS <u>ALL</u> THIS SYNAPSIS HOMOLOGOUS?

In humans, Solari (1980) found that average synapsis between the X and Y extended through the centromere into the long arm and Chandley, et al. (1983) found up to 70% of the Y chromosome synapsed.

A representative sample of the molecular data on this issue has been summarized by Ashley (1984). Briefly, there are many DNA sequences that are unique to the Y that map to the pairing region shared by the X, and many sequences that are common to the X and Y that do not map to corresponding loci in the pairing regions of these chromosomes, as one would predict if all synapsis were homologous. Therefore, both the cytological and molecular data clearly indicate that the answer to this first question is <u>no</u>.

IS <u>ANY</u> OF THIS SYNAPSIS HOMOLOGOUS?

In the last several months there have been several reports of genetic exchange of DNA sequences between the X and Y of normal males. Rouyer, et al. (1986) have compared recombination frequency between three RFLP's common to the X and Y, one of which, DXYS14, is a telomeric sequence (Cooke, et al., 1985). The evidence suggests that these sequences are all located on the distal ends of the short arms. If this DNA is involved in synapsis (and there is no reason to suppose it is not), it seems safe to conclude that some synapsis between the X and Y in man involves homology.

IS THE ATTRACTION BETWEEN THE X AND Y IN SPERMATOCYTES ACTUALLY BASED ON HOMOLOGY?

This must be considered as a separate question from that presented above, since attraction might be based on something other than homology. One test of this issue is provided by XYY mice. If the X and Y recognize homology, the two Y's in XYY males should be at least as effective at forming bivalents as an X and Y. Is this the case? In humans, one of the two Y's is frequently lost in the germ line, but in mice this seldom occurs. E.P. Evans, et al. (1978) reported the frequency of diakinesis-metaphase I associations in four XYY mice and summarized the data from previous reports by them and two other laboratories. They found that XY bivalents were four times as frequent as YY bivalents. A ratio of 2 XY:1 YY would be expected if recognition were based on homology and a Y was equally as effective at recognizing homology as an X.

Given the two Y's are homologous, if there are multiple sites of initiation one might predict that they could initiate homologous synapsis anywhere along their length. If this is true, formation of an XY bivalent should not prevent formation of a trivalent involving the proximal ends of the Y's. Although 16% of the spermatocyte population were trivalents, none were of the type just described. Instead, the second Y was associated by its telomeric end to the junction of the X and the other Y. Since homologous pairing is by twos, neither this fact nor the frequency of association mentioned above is consistent with homology as the sole basis of attraction between the X and Y. Another aspect of their data of possible relevance to this discussion is their observation of a decrease in the trivalents from the early diakinesis to metaphase I population which could be almost entirely accounted for by an increase in the univalent population. If the trivalent involved even one chiasmata, resolution of the association should leave a bivalent and univalent, not three univalents.

The central issue here is the ability of an X or Y to recognize homology. A resolution of this issue would require a study of pachytene spermatocytes. Unfortunately, to date there has been no such report. XYY mice are generally sterile, occur in a very low frequency in the general population, and not identifiable by phenotype. Therefore, data on this point await the fortuitous synaptonemal complex analysis of one of these rare individuals.

IS SYNAPSIS DISTAL TO THE SITE OF RECOMBINATION RESTRICTED TO HOMOLOGY OR IS IT INDIFFERENT TO HOMOLOGY?

The first cytological evidence of a chiasmata between the X and Y came from XY,Sxr (sex reversed) mice. These males produce XX,Sxr mice that are chromosomally female but phenotypically male. They had been an enigma to cytogeneticists until Singh and Jones (1982), were able to show that XY,Sxr males carry a chromosomal aberration in which a male specific DNA segment called Bkm normally present near the centromere of the Y, is duplicated on the distal end of that chromosome Y. They also showed that this distal duplication was present on the distal end of the X in XX,Sxr individuals. E.P. Evans, et al. (1982) were able to cytologically demonstrate a chiasmata in XY,Sxr individuals, a cytological event that has yet to be observed in normal males. If synapsis between the X and Y is homologous, the duplicated Bkm fragment on the distal end of the Y which is recombining with the X and, therefore, obviously lies within the pairing region, should not synapse with the normal X which has none of these sequences. This Bkm fragment has been estimated to consist of a minimum of 150 kb (Jones and Singh, 1981). If synapsis distal to the site of genetic recombination is homologous, one would predict the Bkm sequences should produce a perturbation such as an axial "tail" on the Y, or a "mini-buckle" in the pairing region, if the DNA is inserted into the distal (but not terminal) region. On the other hand, if synapsis of the X and Y is indifferent to homology, one would predict a perfectly normal appearing synaptonemal complex between the X and Y in these animals. We have conducted an electron microscopic examination of pachytene spermatocytes from XY,Sxr males and find normally synapsed XY's. Since nothing is currently known about the minimum size insertion or addition of DNA necessary to produce a disruption in continuity of the synaptonemal complex, failure to detect such a disruption cannot be taken as proof of nonhomologous synapsis. However, the presumed size of the rearrangement is large enough to suggest that caution is warranted in assuming this aberration does not involve nonhomologous synapsis.

THE MOST EXTENSIVE SYNAPSIS BETWEEN THE X AND Y OCCURS DURING EARLY PACHYTENE.

DOES THE TIME COURSE OF NONHOMOLOGOUS SYNAPSIS BETWEEN THE X AND Y FOLLOW THAT OBSERVED IN THE AUTOSOMES?

A particular type of nonhomologous synapsis, termed "synaptic adjustment" has been well characterized. In bivalents heteromorphic for a duplication (Poorman et al., 1981a) or an inversion (Poorman, et al., 1981b) synapsis during early pachytene appeared to be restricted to homology with buckles (duplications) or loops (inversions) in the synaptonemal complex configurations that look like classic textbook diagrams of such aberrations. However, as pachytene progresses the buckle (Moses and Poorman, 1981) or loop (Moses, et al., 1982) "pulls out" and synapses nonhomologously until the aberrant bivalent is indistinguishable morphologically from a normal bivalent. Genetic and cytological data (Moses, et al., 1982), suggest that recombination occurs during the time heteromorphic bivalents are synapsed homologously.

Maximum extent of synapsis of the X and Y in mouse, man, and most other mammals occurs during early pachytene. In fact, extensive synapsis between the X and Y is one of the primary characteristics employed to identify this stage. As discussed above, much of this extensive synapsis must be nonhomologous. Therefore, timing of nonhomologous synapsis of the sex chromosomes does not coincide with timing of the previously defined nonhomologous synapsis of the autosomes. If, as the autosomal data suggest, recombination can occur during early pachytene this early nonhomologous synapsis would appear to increase the risk of nonhomologous exchange between the sex chromosomes.

THERE IS MEIOTIC EXCHANGE OF DNA BETWEEN THE DISTAL TIPS OF THE X AND Y IN MOUSE AND MAN

The evidence for genetic exchange between the X and Y was briefly discussed above and has been presented in much greater detail elsewhere in this symposium. However, this exchange, while resolving a major issue, raises other questions.

IS THIS EXCHANGE EQUIVALENT TO A CROSSOVER AND/OR CHIASMATA IN THE AUTOSOMES?

A structure termed a "recombination nodule" can be found in certain types of pachytene preparations positioned on the synaptonemal complex. They have been found to correspond in frequency and distribution to frequency and distribution of crossovers and chiasma in a number of organisms (Carpenter, 1984 for review).

If recombination nodules are an indicator of recombination, does the <u>frequency</u> of recombination nodules relative to chiasmata on the X and Y correspond to similar data for the autosomes? Rasmussen and Holm (1978) examined recombination nodules in pachytene spermatocytes in humans. For the autosomes they found a close correspondence between the distribution and frequency of recombination nodules and the distribution and frequency of reported chiasma. Every autosome had at least one recombination nodule per bivalent. In a later study, Rasmussen and Holm (1984) reported that the frequency of recombination nodules in early pachytene was higher than the reported chiasma frequency, but that the frequency of these structures by mid-pachytene was very similar to the observed chiasma frequency. These data suggest that there are more recombination nodules present on the synaptonemal complexes in early pachytene than result in detectable chiasma.

Information for the sex chromosomes is quite different. Pooled data from Rasmussen and Holm (1978) and Solari (1980) shows that only 42% of all XY pairs observed in late zygotene and only 57% of all XY pairs observed in early pachytene have recombination nodules. Recombination nodules are not observed on XY pairs after early pachytene. This fact is probably at least partly due to the premature desynapsis of the XY by this stage. However, the evidence clearly shows that there is not a 1:1 correlation between the number of recombination nodules observed on the X and Y and the number of end associations (presumed chiasma) of these chromosomes observed at diakinesis-metaphase I, similar to that observed for autosomes. Moreover, the actual number of recombination nodules observed on the X and Y were measured at a stage when the autosomal information suggests that not all recombination nodules observed during this period will necessarily result in formation of a chiasmata.

Does the distribution of recombination nodules on the X and Y in man correspond to the genetic and cytological data? Solari (1980) reported the average amount of synapsis between the X and Y to be around 1.4 microns or 30% of the Y. Based on calculations of the length of the short arm of the Y, this average synapsis extends through the kinetochore region. Rouyer, et al. (1986) report that 90% of the DNA sequences recovered to date that map to Yp are unique and do not map to the distal end of Xp. Presumably this means that no more than approximately 10% of Yp can be synapsing homologously with the X. Solari (1980) found only 9 early pachytene spermatocytes with recombination nodules, six of which were located in the distal .5 micron of the pairing segment. Even though this segment is one third to one half of the short arm of the Y, since the exact positions of these nodules were not given, it could be assumed that their distribution was not inconsistent with involvement in a recombination event in the presumptive homologously paired region. However, he reports three recombination nodules in the next .5 micron segment. This region extends almost to the centromere. Therefore it is highly likely a recombination event at these sites could be expected to result in exchange of nonhomologous sequences. Since there is no evidence of this high a frequency of nonhomologous exchange, the most logical intrepretation of the data is not all recombination nodules observed on the X and Y actually result in exhange.

This leads to a crucial question: "Is nonhomologous exchange between the X and Y prohibited?"

XX males occur with a frequency of about 1 per 20,000 in the general population (Bishop, et al., 1985). Although many of these individuals are XX/XY mosaics, nonmosaic XX males exist. Bishop, et al. (1985) have recently probed the DNA of a number of these individuals for Y specific sequences. They found that 66% of these individuals contained varying numbers of Y specific fragments which could be shown to be located on the X chromosome. These data along with the cytological evidence of H. J. Evans, et al. (1979) suggest that nonhomologous exchange not only <u>can</u> occur, but <u>does</u> occur within the human population.

WHAT PREVENTS MORE FREQUENT EXCHANGE BETWEEN THE EXTENSIVELY NONHOMOLOGOUSLY SYNAPSED X AND Y?

Premature desynapsis of the nonhomologously synapsed regions has been suggested as a possible mechanism for prevention of recombination in this region (Ashley and Moses, 1980). This idea is viable if recombination can only occur <u>after</u> the X and Y have desynapsed. Although the exact time of crossover is not currently known, the data discussed above suggest that it may occur during early to mid-pachytene. If this is indeed the case, premature desynapsis may be insufficient to prevent nonhomologous exchange.

Jones (1985) has proposed that the conformation of the Y chromatin itself may normally prevent recombination. This theory is attractive since it is time independent. However, it also has a major drawback. The prime Y chromatin candidate proposed for this role is the Bkm fragment. As discussed above, XY,Sxr is a chromosomal rearrangement that has apparently resulted in a duplication of the Bkm fragment at the distal end of the Y in mice. If the conformation of this sequence normally prevents crossing over, the rearrangement has resulted in an alteration of the conformation so that crossover is no longer prevented.

ARE THE SEX CHROMOSOMES INVOLVED IN ABERRANT EXCHANGES NORMAL, OR, ARE THEY THEMSELVES CHROMOSOMAL REARRANGMENTS?

Aberrant exchange between the X and Y could be the result of an alteration in the cellular physiology of the spermatocytes involved, leading to an exchange when and where it does not ordinarily occur. Alternatively it could be the result of an alteration in the chromosomes themselves. The first evidence of recombination between the X and Y came from XY,Sxr mice. Sex reversal in this case unequivocally occurs in a rearranged Y chromosome. In this light and in the context I have discussed above, I suggest that the parents of XX males should be examined for evidence of rearranged sex chromosomes.

IN THOSE MAMMALS IN WHICH THERE IS NO SYNAPSIS BETWEEN THE X AND Y, WHAT CYTOLOGICAL MECHANISM ASSURES THEIR NORMAL DISJUNCTION?

As mentioned above, there are mammals for which there is no evidence of synapsis between the X and Y during any stage of meiotic prophase. In the marsupials, the ends of the X and Y are joined at the site of a modified nuclear membrane. In other mammals in which there is no synapsis, there is no evidence of a special structure, but there appears to be a physical attachment between the ends of the X and Y (Solari and Ashley, 1977). The nature of this attachment is not known, but a possible DNA connection has never been tested. It would be of interest to determine whether a physical exchange of DNA similar in nature to the exchange of telomeric sequence exchange observed in man (Cooke, 1985) occurs between the X and Y telomeric sequences of these species.

Whatever, the outcome of such experiments, it currently appears inappropriate to speak of homologous synapsis and genetic exchange involving recombination nodules and chiasmata formation as a universal mechanism of disjunction of sex chromosomes.

SUMMARY

The meiotic behavior of the sex chromosomes appear to be providing us with the opportunity to examine variations on the meiotic theme, as well as to ask what is happening in some clinically abnormal situations. These situations are difficult to explain by the Darlington hypothesis, but may be more readily understandable in the context discussed here. It would be ill advised for us to become so involved in proving that the X and Y fit the conventional autosomal model of disjunction that we miss the opportunity to gain an increased understanding of the entire spectrum of events.

ACKNOWLEDGEMENTS

This work was supported by NIH grant HD 18470.

References

Ashley, T. 1984. A re-examination of the case for homology between the X and Y chromosomes of mouse and man. Hum. Genet. 67: 372-377.

Ashley, T., and M.J. Moses. 1980. End congression and segregation of the achiasmatic sex chromosomes of the sand rat (Psammomys obesus). J. Cell Sci. 50: 105-119.

Bishop, C., J. Weissenbach, M. Casanova, A. Bernheim, and M. Fellows. 1985. DNA sequences and analysis of the human Y chromosome. In: The Y Chromosome, Part A: Basic Characteristics of the Y Chromosome, p. 141-176. A.A. Sandberg, ed. Alan R. Liss.

Carpenter, A.T.C. 1984. Recombination nodules and the mechanism of crossing-over in Drosophila. In: Controlling Events in Meiosis. C.W. Evans and H.G. Dickinson eds. pp. 233-243.

Chandley, A.C., P. Goetz, P. Hargreave, A.M. Joseph, and R.M. Speed. 1984. On the nature and extent of XY pairing at meiotic prophase in man. Cytogenet. Cell Genet. 38: 241-247.

Cooke, H.J., W.R.A. Brown, and G.A. Rappold. 1985. Hypervariable telomeric sequences from the human sex chromosomes are pseudoautosomal. Nature 317: 687-691.

Evans, E.P., C.V. Beechey, and M.D. Burtenshaw. 1978. Meiosis and fertility in XYY mice. Cytogenet. Cell Genet. 20: 249-263.

Evans, E.P., M.D. Burtenshaw, and B.M. Cattanach. 1982. Cytological evidence for meiotic crossing over between the X and Y chromosomes in male mice carrying the sex reversing (Sxr) factor. Nature 300: 443-445.

Evans, H.J., K.E. Buckton, G. Spowart, and A.D. Carothers. 1979. Heteromorphic X chromosomes in 46,XX males: evidence for the involvement of X-Y interchange. Hum. Genet. 49: 11-31.

Jones, K.W. 1985. Bkm Sequences and their conservation. In: The Origin and Evolution of Sex. H.O. Halvorson and A. Monroy pp. 199-211. Alan R. Liss.

Jones, K.W., and L. Singh. 1981. Conserved repeated DNA sequences in vertebrate sex chromosomes. Hum. Genet. 58: 46-53.

Koller, P.C., and C.D. Darlington. 1934. The genetical and mechanical properties of the sex chromosomes. I. Rattus norvegicus male. J. Genet. 29: 159-173.

Moses, M.J., and P.A. Poorman. 1981. Synaptonemal complex analysis of mouse chromosomal rearrangements. II. Synaptic adjustment in a tandem duplication. Chromosoma 81: 519-535.

Moses, M.J., P.A. Poorman, T.H. Roderick, and M.T. Davisson. 1982. Synaptonemal complex analysis of mouse chromosome rearrangements. IV. Synapsis and synaptic adjustment in two paracentric inversions. Chromosoma 84: 457-474.

Pathak, S., F.B. Elder, and B.L. Maxwell. 1980. Asynaptic behavior of the X and Y in the Virginia opossum and the southern pygmy mouse. Cytogenet. Cell Genet. 26: 142-149.

Poorman, P.A., M.J. Moses, M.T. Davisson, T.H. Roderick. 1981b. Synaptonemal complex analysis of mouse chromosome rearrangements. III. Cytogenetic observations on two paracentric inversions. Chromosoma 83: 419-429.

Poorman, P.A., M.J. Moses, L.B. Russell, and N.L.A. Cacherio. 1981a. Synaptonemal complex analysis of mouse chromosomal rearrangements. III. Cytogenetic observations on a tandem duplication. Chromosoma 81: 419-429.

Rasmussen, S.W., and P.B. Holm. 1978. Human Meiosis II. Chromosome pairing and recombination nodules in human spermatocytes. Carlsberg Res. Commun. 4: 275-327.

Rasmussen, S.W., and P.B. Holm. 1984. The synaptonemal complex, recombination nodules and chiasmata in human spermatocytes. In: Controlling Events in Meiosis. C.W. Evans and H.G. Dickenson eds. Company of Biologists. Cambridge. pp. 271-292.

Rouyer, F., M.-C. Simmler, C. Johnsson, G. Vergnaud, H.J. Cooke, and J. Weissenbach. 1986. A gradient of sex linkage in the pseudoautosomal region of the human sex chromosomes. Nature 319: 291-295.

Sharp, P. 1982. Sex chromosome pairing during male meiosis in marsupials. Chromosoma 86: 27-47.

Singh, L., and K.W. Jones 1982. Sex reversal in the mouse (Mus musculus) is caused by a recurrent non-reciprocal crossover involving the X and an aberrant Y chromosome. Cell 36: 111-120.

Solari, A.J. 1980. Synaptonemal complexes and associated structures in microspread spermatocytes. Chromosoma 81: 315-337.

Solari, A.J., and T. Ashley. 1977. Ultrastructural and behavior of the achiasmatic, telosynaptic XY pair of the sand rat Psammomys obesus. Chromosoma 62: 319-336.

Tres, L.L. 1977. Extensive pairing of the XY bivalent in mouse spermatocytes as visualized by whole-mount electron microscopy. J. Cell Sci. 25: 1-15.

DNAse I hypersensitivity characterizes the XY pairing region at meiosis in man

A. C. Chandley and S. McBeath

MRC Clinical and Population Cytogenetics Unit, Western General Hospital, Edinburgh EH4 2XU, Scotland

INTRODUCTION

Active or potentially active genes at interphase are organized in chromatin in a conformation that is distinct from that of inactive genes, as illustrated by their sensitivity to digestion by DNAse I (see Weisbrod 1982 for review). Mitotic metaphase chromosomes retain this characteristic and DNAse I sensitive regions can be visualized by applying nick translation and autoradiography to fixed chromosome spreads (Gazit et al 1982). DNAse I sensitive regions then show a close correspondence with light G-bands (Kerem et al 1984) which tend to house expressed genes (Kerem et al 1983; Kuo and Plunkett 1985) and undergo early replication (Goldman et al 1984). In female mitotic cells, only those regions on the inactive X chromosome which are early replicating show a positive response to the nick translation procedures, these being regions which escape inactivation (Kerem et al 1983; Sperling and Marcus 1984; Sperling et al 1985).

We have investigated the response of the human XY bivalent at metaphase I to nick translation using DNAse I and autoradiography. Our findings show a differential sensitivity to the nicking enzyme with specific sites being preferentially labelled. The Xpter;Ypter pairing region is among the most hypersensitive sites. A change in chromatin conformation in this region from mitotic to meiotic cells is suggested, this perhaps being a prerequisite for crossing-over of sequences between the sex chromosomes.

METHODS

i Slide Preparation

Meiotic air-dried preparations (Evans et al 1964) of four patients (WSM 195, 199, 206, 208), were used for the experiments. All were chromosomally normal and showed a normal spermatogenic picture. Data were obtained from five runs of the experiment (Runs 1, 4, 6, 7 and 8). Runs 7 and 8 employed the use of blood lymphocyte chromosome suspensions air-dried on the same slides as meiotic suspensions from the same individuals. Direct comparisons between mitotic and meiotic metaphases in response to NT incubation were thus possible for two different patients. Nick translation/autoradiographic procedures were as described by Kerem et al (1984). Control slides were prepared by omitting DNAse I from the mixture. NT incubations of 5', 10' and 30' were used at various times and in different experiments. Slides were rinsed at room temperature through a graded series of alcohols (70% to 95% to absolute ethanol), washed for 5' in 5% TCA at 4°C, then rinsed again through the alcohols. Dried slides were stained with carbol fuchsin (Carr and Walker 1961) prior to filming. Autoradiographs were prepared using Ilford L4 liquid emulsion. Exposure periods of from 1 day to 28 days at 4°C were given. Kodak D19 developer was used.

ii Chromosome identification

Identification of the human XY bivalent at metaphase I of meiosis in carbol fuchsin stained preparations is clear and unambiguous owing to the chain-like configuration formed by the end-to-end pairing of the X and Y short arm tips. For unambiguous orientation of the XY bivalent, sequential staining using Distamycin A (DA)/DAPI (Schweizer et al 1978) was applied after removal of the carbol fuchsin stain. This gave fluorescence at the Yq heterochromatin and to a lesser extent over the X and Y centromeres, thus providing landmarks for identification. Representative cells showing specific sites or "hot-spots" of labelling over the X and Y chromosomes were photographed before and after fluorescence staining.

For sex chromosome labelling comparisons between meiotic and mitotic cells, the Y chromosome only was used as this can usually be identified in blood metaphases by morphological criteria alone. Grain count comparisons with univalent Y chromosomes at MI were made.

RESULTS

i Patterns of labelling seen over the XY bivalent of DNAse I treated cells

All clear well-spread metaphase I (MI) divisions from WSM 195 (Run 1) and WSM 199 (Run 4), were systematically analysed in 9 NT-treated (5', 10' and 30' incubations) and 2 control slides. All slides were exposed for 1 day to autoradiographic emulsion. It was immediately observed that a high proportion of MIs showed localised grain clusters or "hot-spots" of label against an otherwise largely unlabelled sex bivalent, particularly at the 5' and 10' NT incubation intervals (Table).

TABLE Frequency of cells showing label hot-spots over the XY bivalent at meiotic metaphase I in DNAse I treated (1ng/ml at 15°C) and control cells. In all experiments, a 24 hour autoradiographic exposure was given.

			% Distribution of labelling patterns seen over XY bivalents		
	Nick translation treatment group	No of cells analysed	Hot spots	Random distribution	Too few grains
RUN 1 (WSM 195)	5'	25	44	0	56
	10'	30	63	7	30
RUN 4 (WSM 199)	Control	63	8	0	92
	10'	70	57	17	26
	30'	100	7	93	0

At these treatment times, from 44% to 63% of all MIs observed, showed hot-spot labelling, with clusters of from 2 to 7 grains occurring over one or more specific sites. The remaining cells had either too few grains to discern localizations, or were somewhat randomly labelled. Following a 30' NT incubation, the proportion of MIs showing hot-spots decreased to control levels as heavier and more random labelling occurred along the XY bivalent.

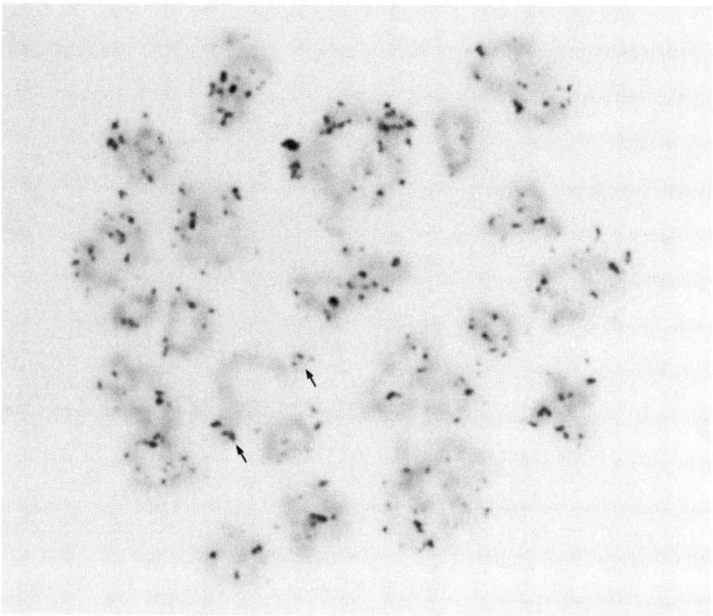

FIGURE 1 Meiotic metaphase I (MI) from patient WSM 199 (Run 4) showing grain clusters over the terminal/telomeric regions of Xq and Yq (arrows). Air-dried preparation stained carbol fuchsin. 10' nick translation (NT) incubation. 24hr autoradiographic exposure (ARG).

Three principle sites of clustered labelling were noted when carbol fuchsin preparations were examined. About half of all cells with hot-spots showed grain localization over the Xq and/or Yq terminal (sometimes telomeric) segments (Figure 1). The other half showed a cluster of grains about one-third of the way along the XY bivalent over what appeared in some cells to be the XpYp pairing region. Sequential staining with DA/DAPI confirmed this location for the majority (Figures 2a and b), but in a minority of cells, this labelled site was deduced to be at a position just below the centromere of the X, in the long arm (Figures 3a and b).

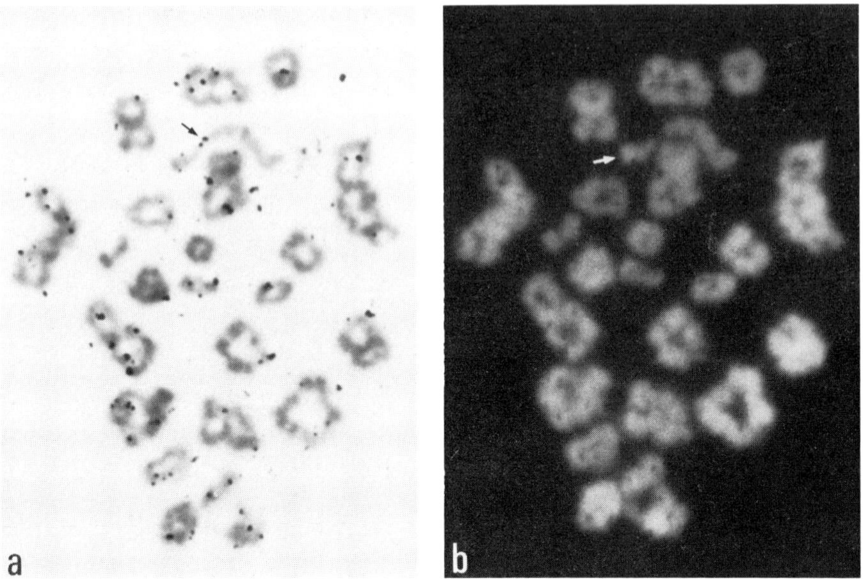

FIGURE 2 MI from patient WSM 195 (Run 1) showing a a grain cluster (arrow) along the XY bivalent in a carbol fuchsin stained cell. In b the same cell stained DA/DAPI shows the fluorescent Y chromosome (arrow) giving confirmation of grain localisation over the Xpter;Ypter pairing region (5' NT incubation;24hr ARG).

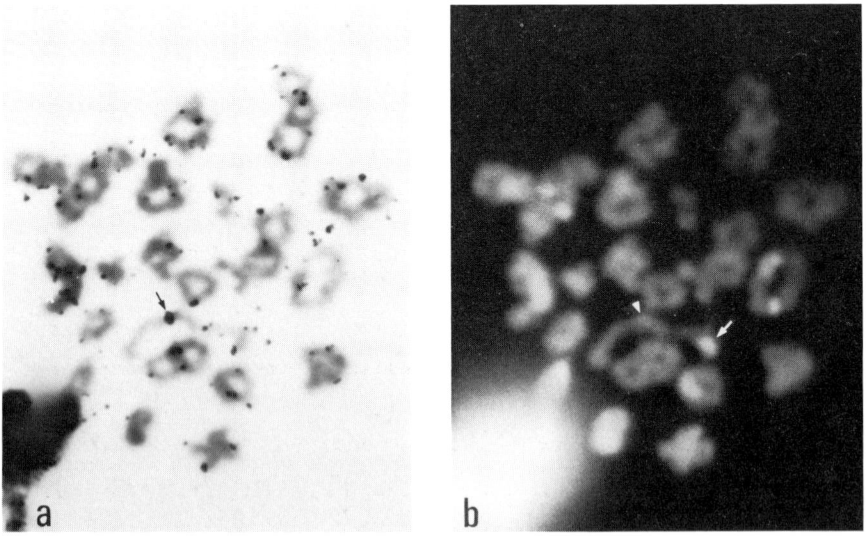

FIGURE 3 MI from patient WSM 195 (Run 1) showing a grain localisation along the XY bivalent (arrow). In b the fluorescent Y chromosome (arrow) and X centromere (arrowhead) provide landmarks to localize the grain cluster to a site just below the centromere in the long arm of the X. (10' NT incubation;24hr ARC).

For all the many cells showing hot-spots of label along the XY bivalent or on X and Y univalents, the theme was repeated: clusters appeared over one or two of the above-mentioned sites, or in a few cases over three or even four sites independently. Furthermore, the same patterns of XY labelling were preserved in cells given similar short NT incubations but longer periods of autoradiographic exposure, the sites and cells in general then, however, being much more heavily labelled.

ii Patterns of labelling seen over control cells

Control cells given a one day exposure to autoradiographic film were very lightly labelled (Figure 4). For patients WSM 206 (Run 7) and WSM 208 (Run 8), one control slide was left exposing to film at 4°C for 28 days. At the end of this time, heavy

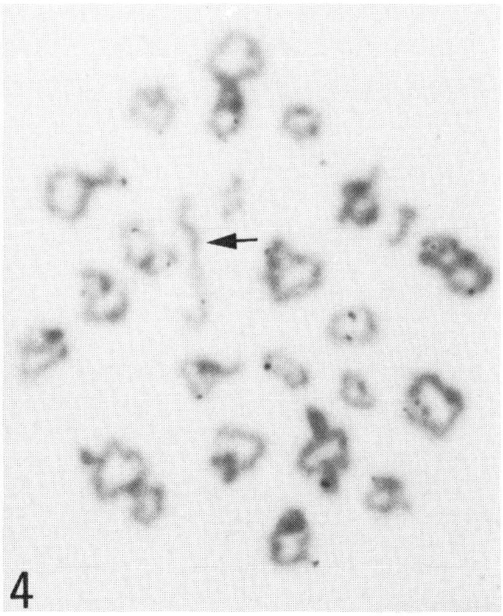

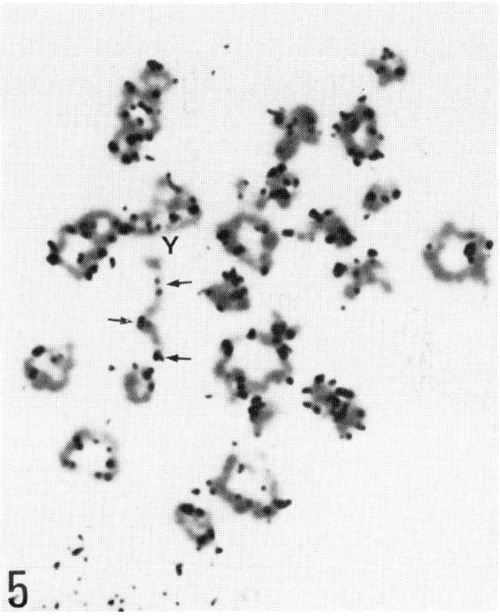

FIGURE 4 Control MI from patient WSM 199 (Run 4). The XY bivalent (arrowed) is virtually unlabelled as are the autosomal bivalents (24h ARG).

FIGURE 5 Control MI from patient WSM 206 (Run 7) given a 28-day exposure to autoradiographic film. Label hot-spots can be seen over the Xpter;Ypter pairing region, at the X centromere and at Xqter (arrows).

deposition of grains was seen over all autosomal bivalents and over the XY bivalent. In occasional cells, however, hot-spots of label were seen over a largely unlabelled XY pair. These sites mimicked those described above for DNAse I treated cells given a 1 day exposure (Figure 5). The effects of fixatives and other pretreatments given to cells prepared by the air-drying method could account for this non-DNAse I induced nicking of the DNA. Y chromosome label hot-spots were not found for mitotic metaphases given a one month exposure to film although occasional telomeric clusters over autosomes were seen. In general, however, label was distributed also over many other non-telomeric sites.

iii Grain comparisons for the Y chromosome in mitotic and meiotic metaphases

In Runs 7 and 8, grain count estimations were made for twenty metaphases and for the Y chromosome alone in blood, and for twenty MIs and for the Y chromosome when univalent at meiosis. A full account of the data obtained will be reported elsewhere (Chandley and McBeath, in preparation). Suffice it to say here that in general, counts for whole cells were higher at meiosis than in mitosis by a factor of 1-2 times. For the Y chromosome alone, levels of labelling were 3-4 times higher at meiosis (MI), mainly due to the occurrence of hot-spot labelling. Hot-spots of label were never seen over mitotic Y chromosomes under conditions comparable to those which revealed such sites at meiosis.

DISCUSSION

Our results provide clear evidence of differential sensitivity to DNAse I along the human XY bivalent at metaphase I of meiosis, as demonstrated by nick translation and autoradiography. Under optimized experimental conditions, grain clustering is observed at a limited number of sites on the sex chromosomes at meiosis, a clustering which has not been observed at mitosis, at least for the Y chromosome, under comparable conditions of nick translation. Sites of hypersensitivity to DNAse I include the Xpter;Ypter pairing region, the terminal and/or telomeric regions of Xq and Yq, and a site just below the centromere in Xq. In all of these restricted regions, it would appear that a DNA conformation exists at meiosis which preferentially permits accessibility to the nicking enzyme, the XY bivalent for the most part remaining unlabelled.

The finding in control meiotic cells that these same DNA segments are accessible to artefactual nicking reinforces the idea of a specific conformational difference between the nick-translated and non-nicked portions of the XY chromatin.

In general, the sex chromosomes at meiosis in the male are condensed and transcriptionally inactive (Monesi 1967; Tres 1975), and this is reflected in the findings of lower overall levels of labelling for the XY pair compared with the autosomes, not only in our experiments with human spermatocytes, but in earlier nick translation experiments which have been carried out using the mouse (Sperling and Marcus 1984). Hot-spots of differential sensitivity along the XY bivalent, similar to those seen in man, were not, however, revealed under the conditions used in these experiments with the mouse. Nick translation generally correlates with gene activity or potential activity, and tends to occur preferentially over early replicating segments of DNA (Kerem et al 1983, 1984). Replication timings for specific bands in meiotic chromosomes are not available, but for the human X and Y chromosomes at mitosis, BrdU-Hoechst replication studies have shown that among the earliest replicating regions in the genome are several bands which cover the XpYp pairing region at meiosis, and band Xq 13.1, a site just below the centromere in the X long arm (Schempp and Meer 1983). The latter has been postulated also to be the human X inactivation centre around which the Barr body in the female condenses (Therman et al 1979). Both the pairing region between the X and Y and a site which could be comparable to band Xq 13.1 at mitosis, show hypersensitivity to DNAse I attack in our experiments using spermatocytes, supporting the general correlation that exists between early replication and nick translation. The pairing tips of the X and Y chromosomes at meiosis in man display an additional set of interesting features. They have shared homology between several DNA sequences which are highly polymorphic and which undergo exchange with a high frequency, the rate of exchange increasing with proximity to the telomere (Rouyer et al 1986).

It is of interest that label over the Xq and Yq terminal regions in our experiments seemed occasionally to be localised over the extreme tips of the chromosome arms, suggesting telomeric hypersensitivity to DNAse I. Telomeric sensitivity both to enzyme attack and control nicking could come about through the peculiar characteristics of the DNA at the ends of chromosomes, consisting as it does of simple satellite-like tandemly repeated sequences containing single-strand gaps at specific positions within the cluster of repeats (see Blackburn and Szostak 1984 for review). It is of perhaps some significance that pairing between the Xq and Yq tips is a common observation at prophase of meiosis in man (Chandley et al 1984), although such associations are usually lost by the time cells have progressed to metaphase I. The special conformational nature of the long arm tips, as revealed by their sensitivity to DNAse I attack, could facilitate such pairing. When labelling is observed over the distal segment of the long arm of the X, it could relate also to the special sensitivity of bands Xq 27-28, a known site of increased recombination in females (Szabo et al 1984; Oberle et al 1985), the DNA again showing a high degree of polymorphism (Oberle et al 1985). In individuals with a fragile site (Sutherland 1979) this region is also one showing vulnerability to breakage.

Our DNAse I treatments of the X and Y chromosomes at meiosis could therefore be revealing sites where recombinational hot-spots occur in the human genome. Indeed, the chromatin conformational changes which are suggested by our experiments to have occurred in these regions between mitosis and meiosis might be an essential prerequisite for crossing-over. Such a switch in DNA conformation from somatic to germ-line cells would not be unprecedented. Raman and Nanda (1986) have shown in the shrew, Suncus murinus, that the region of chromatin in which a chiasma forms at meiosis, changes at pachytene from being C-band positive and condensed to being C-band negative and diffuse, and they too have speculated on the possibility that this conformational change is a prerequisite for crossing-over. This same change from C-band positive to C-band negative has been observed in meiocytes of the Chinese hamster (Murer-Orlando and Richer 1983) and Armenian hamster (Lavappa 1983), two other species in which a chiasma is formed between the X and Y in males. In the Chinese hamster, the heterochromatin within the pairing region which allows the formation of a chiasma, is of a particular nature, showing lower resistance to alkali treatment, and a lack of repetitive DNA sequences. Interestingly enough, a principle site of gene expression and nick translation in the Chinese hamster has been demonstrated at the telomere of the Y short arm in somatic cells (Kuo and Plunkett 1985), and it is by this end of the Y chromosome that pairing takes place at meiosis with the X (Murer-Orlando and Richer 1983). In several species, including the Chinese hamster showing chiasma formation between the X and Y at meiosis, the region involved in crossing-over also tends to complete its replication before the remainder of the Y chromosome (Raman and Nanda 1986). A hypothesis linking meiotic pairing and recombination with early replication has recently been proposed (Chandley 1986).

Early replication and DNAse I sensitivity shown by nick translation, perhaps owing to the opening up of chromatin to allow for recombination, thus appear to correlate for the XY pairing region in the limited number of species investigated so far. It may well turn out to be, when more species have been examined, that XY pairing regions universally will be found to share these common properties and characteristics. For autosomes too, a more open conformation might characterise sites where chiasmata are formed at meiosis and this may account for the higher levels of DNAse I sensitivity shown in our experiments between mitotic and meiotic metaphases. Further detailed investigation of specific autosomal bivalents will be required to establish whether this speculation holds true.

REFERENCES

Blackburn, E. H. and Szostak, J. W. 1984. The molecular structure of centromeres and telomeres. Ann. Rev. Biochem. 53, 163-94.

Carr, D. H. and Walker, J. E. 1961. Carbol fuchsin as a stain for mammalian chromosomes. Stain Technol. 36, 233-36.

Chandley, A. C. 1986. A model for effective pairing and recombination at meiosis based on early replicating sites (R-bands) along chromosomes. Hum. Genet. 72, 50-7.

Chandley, A. C., Goetz, P., Hargreave, T. B., Joseph, A. M. and Speed, R. M. 1984. On the nature and extent of XY pairing at meiotic prophase in man. Cytogenet. Cell Genet. 38, 241-47.

Evans, E. P., Breckon, G. and Ford, C. E. 1964. An air-drying method for meiotic preparations from mammalian testes. Cytogenetics 3, 289-94.

Gazit, B., Cedar, H., Lerer, I. and Voss, R. 1982. Active genes are sensitive to deoxyribonuclease I during metaphase. Science 217, 648-50.

Goldman, M. A., Holmquist, G. P., Gray, M. C., Caston, L. A. and Nag, A. 1984. Replication timing of mammalian genes and middle repetitive sequences. Science 224, 686-92.

Kerem, B., Goitein, R., Diamond, G., Cedar, H. and Marcus, M. 1984. Mapping of DNAse I sensitive regions on mitotic chromosomes. Cell 38, 493-99.

Kerem, B., Goitein, R., Richler, C., Marcus, M. and Cedar, H. 1983. In situ nick translation distinguishes between active and inactive X chromosomes. Nature 304, 88-90.

Kuo, M. T. and Plunkett, W. 1985. Nick-translation of metaphase chromosomes: In vitro labelling of nuclease-hypersensitive regions in chromosomes. Proc. Nat. Acad. Sci. 82, 854-58.

Lavappa, K. S. 1983. Heterochromatin in X and Y chromosomes of the Armenian hamster. XVth Int. Cong. Genet. New Delhi 155 (Abstract).

Monesi, V. 1967. Ribonucleic acid and protein synthesis during differentiation of male germ cells in the mouse. Arch. Anat. Microscop. Morphol. Experimentale 56, 61-74.

Murer-Orlando, M. and Richer, C-L. 1983. Heterochromatin heterogeneity in Chinese hamster sex bivalents. Cytogenet. Cell Genet. 35, 195-99.

Oberle, I., Drayna, D., Camerino, G., White, R. and Mandel, J-L. 1985. The telomeric region of the human X chromosome long arm: Presence of a highly polymorphic DNA marker and analysis of recombination frequency. Proc. Natl. Acad. Sci. 82, 2824-28.

Raman, R. and Nanda, I. 1986. Mammalian sex chromosomes. I. Cytological changes in the chiasmatic sex chromosomes of the male musk shrew Suncus murinus. Chromosoma 93, 367-74.

Rouyer, F., Simmler, M-C., Johnsson, C., Vergnaud, G., Cooke, H.J. and Weissenbach, J. 1986. A gradient of sex linkage in the pseudoautosomal region of the human sex chromosomes. Nature 319, 291-95.

Schempp, W. and Meer, B. 1983. Cytologic evidence for three human X-chromosomal segments escaping inactivation. Hum. Genet. 63, 171-4.

Schweizer, D., Ambros, P. and Andrle, M. 1978. Modification of DAPI banding on human chromosomes by prestaining with a DNA-binding oligopeptide antibiotic, distamycin A. Exp. Cell Res. 111, 327-32.

Sperling, K. and Marcus, M. 1984. Mapping of gene activity on mammalian chromosomes. In Chromosomes Today, vol. 8, 169-78. George Allen and Unwin.

Sperling, K., Kerem, B., Goitein, R., Kottusch, V., Cedar, H. and Marcus, M. 1985. DNAse I sensitivity in facultative and constitutive heterochromatin. Chromosoma 93, 38-42.

Sutherland, G. R. 1979. Heritable fragile sites on human chromosomes. Hum. Genet. 53, 23-7.

Szabo, P., Purrello, M., Rocchi, M., Archidiacono, N., Alhadeff, B, Filippi, G., Toniolo, D., Martini, G., Luzzatto, L. and Siniscalco, M. 1984. Cytological mapping of the human glucose 6-phosphate dehydrogenase gene distal to the fragile-X site suggests a high rate of meiotic recombination across this site. PNAS 81, 7855-59.

Therman, E., Sarto, G. E., Palmer, C. G., Kallio, H. and Denniston, C. 1979. Position of the human inactivation center on Xq. Hum. Genet. 50, 59-64.

Tres, L. 1975. Nucleolar RNA synthesis of meiotic prophase spermatocytes in the human testis. Chromosoma 53, 141-51.

Weisbrod, S. 1982. Active chromatin. Nature 297, 289-95.

Chromosome behaviour at female meiosis in two murine (Robertsonian) trisomies*

P. E. Polani and J. A. Crolla

Paediatric Research Unit, United Medical and Dental Schools of Guy's and St Thomas's Hospitals, Guy's Tower, London SE1 9RT, England

1. INTRODUCTION

It is one of the established tenets of classical cytogenetics (Newton and Darlington, 1929; see e.g. Swanson et al., 1981) that pairing during the meiotic prophase of three homologous chromosomes follows the basic pattern characteristic of pairing of two homologues when synapsis establishes bivalents: namely pairing is of two by two threads and never are all three threads paired at any point of their association. Complete belief in the universality of this principle was shaken when Comings and Okada (1971) observed by electron-microscopy (EM) that triple pairing with double, side by side, tripartite synaptonemal complexes (SC) could occur during meiotic pairing in the triploid chicken. Among mammals, the behaviour of trisomic chromosomes at first meiotic metaphase (MI), has been studied in human trisomy (Ts) 21 males (summarised by Jagiello, 1981) and there are also data on chiasma counts in trisomic men (Hultén and Lindsten, 1970) and in one cell from a female (Polani et al., 1982). In mice, the low spontaneous rate of Ts has precluded such work.

The present study combines two experimental techniques, the first of which generates specific murine autosomal trisomies in large numbers (Gropp et al., 1975). As these are almost always embryonic lethals (Gropp, 1976), precluding all but female prophase

* The authors dedicate this paper to the memory of Alfred Gropp

Chromosomes today, volume 9 Allen & Unwin 1987

studies in some Ts, we decided to rescue the fetal ovaries from Robertsonian (Rb) trisomic stock, using the in vitro/in vivo method developed at Guy's (Polani et al., 1979). This approach allowed us to study sequentially both prophase and metaphase cells for analysis of chromosome behaviour, and our preliminary results are presented here.

2. MATERIALS AND METHODS

Rb heterozygous mice with monobrachial homology (for autosome 16 or 19) were mated to all acrocentric stock, A/Strong generating balanced (Bal) 16 or 19, mono-Rb fetuses: (2n = 39; NF = 40) or trisomic, di-Rb, non-disjunctional Ts16 or 19 fetuses: (2n = 39; NF = 41) depending on the Rb heterozygote used. The products of conception were dissected between day (d) 14-19 of gestation and ovaries from trisomic and Bal fetuses were set up in organ culture for up to eight days. Meiotic prophase studies on pachytene oocytes were performed on ovaries explanted at d14 with each 24h period in vitro assumed equivalent to one day in fetu. Other ovaries were cultured for a variable period allowing the germ cells to reach the dictyotene stage of meiosis, and then transplanted into the heterotopic renal site of previously spayed Rb females of the appropriate Ts producing system. For the EM prophase study, cultured ovaries, corresponding to gestational ages of d18 (Ts16) and d19 (Ts19) were processed using the SC surface-spreading method developed by Moses et al., 1982. Mature ova were collected from successfully matured heterotropic transplants 19-21 days after grafting (day of transplant = d1) and cultured at 37°C for either 4h for MI or 16h for MII in Ts16, or 40h for Ts19. Chromosome preparations were made following an air-drying protocol (Polani and Jagiello, 1976) and orcein stained. Measurements of the lengths of chromosomal axes (using a map measurer) were made from photographic enlargements of EM. The distances between axes in double or triple associations were calculated using an eyepiece graticule on a low-power microscope.

3. RESULTS

Following general remarks on the trisomy breeding systems, our observations concern firstly, the manner of pairing of chromosomal axes at the prophase, and therefore, SC-formation; secondly, the way chromosomes are associated at diakinesis and MI, and thirdly, the way MI chromosomes separate as judged from an analysis of MII. A. General. Out of 2623 viable fetuses from both systems, almost one fifth (477) were trisomic. We transplanted over 150 pairs of ovaries (Ts16 and Ts19 combined) with an over 50% success rate yielding on average, 15.5 eggs per successful transplant in Ts16, and 30.2 per successful transplant from the Ts19 systems. B. Analysis of trisomic prophase oocytes (EM). The prophase was sub-staged using standard criteria (Moses and Poorman, 1984). In the Ts16 system late pachynema was the dominant stage on d18, but was seen one day later in Ts19. The observations on pairing derives from a study of 34 oocytes from Ts16 and 63 from Ts19 animals. Pairing of the acrocentric axes resulted in the formation of 17 regular bivalents together with the "Rb complex" of five chromosomes showing different pairing arrangements in different cells (Table I).

TABLE I

Rb Complex: Pairing at meiotic prophase (EM) and configurations at MI (V = pentavalents, etc.) in the two trisomies. Total number of cells (N) (see text).

			Percentages			
		N	V	III + II	II + II + I	IV + I
Ts16	EM	34	79	12	3	6
	MI	97	75	14	1	9
Ts19	EM	63	55	28	7	10
	MI	245	45	40	4	11

In Ts16 practically 80% of Rb complexes formed a pentavalent which involved variable pairing of the two Rb by their homologous arms to which the axis of the acrocentric 16 was also paired, mostly forming triple associations. The non-homologous arms of the two Rb are paired with the axes of their corresponding acrocentrics to form a double bivalent region of the pentavalent. In Ts19, pentavalents are only seen in half the oocytes. The next commonest configuration of the Rb complex is that of an Rb trivalent plus an Rb bivalent, and in the latter, the homologous arm (autosome 16 or 19) is unpaired and thickened. Overall, univalence was found in about 20% Ts16 and in over 40% Ts19 oocytes. In about 90% of oocytes in both Ts, the trisomic axes in pentavalents are triply associated to a variable extent, mostly in the distal region. Measurements showed that the proportion of triple associations was greater in Ts16 than in Ts19, and that the three closely paired axes in the trisomic region were as close to each other, in pairs, as were the axes in the bivalent regions of the two Rb and in the acrocentric bivalents. C. Analysis of MI. **(a.) Distribution of Rb multivalents.** 97 Ts16 and 245 Ts19 MI, were complete and well spread. The Rb complex formed a pentavalent in 75% of Ts16 and in 45% of Ts19 MI (Table I). The acrocentric 16, or 19, was a univalent in 10% and 15% respectively, so that univalence overall was present in 24% Ts16 and, 55% Ts19 MI. **(b.) Analysis of chiasmata (Xma)** Total Xma count in complete cells. In only 12 Ts16 and 21 Ts19 MI could the Xma number and positions in all bivalents and configurations be confidently analyzed, together with a small number of 'complete' cells of the two respective Bal MI. There were an average 29.7 Xma in Ts16 and 25.25 in Bal16 cells and in the Ts19 system the corresponding values were 29.05 and 26.20. Xma in acrocentrics and Rb multivalents. The average number of Xma per bivalent in Ts16 was 1.52 (700 bivalents) compared with 1.30 in Bal16 (448 bivalents); 1.51 in Ts19 (915 bivalents) and 1.37 (728 bivalents) in Bal19 cells. The average number of Xma per Rb configuration decreased as the pentavalents split into lower order configurations (from 5.0 Xma in Ts16 and 4.56 Xma in Ts19 pentavalents to 1.0 Xma in Ts16 and 1.27 Xma in Ts19 Rb bivalents). Balanced and trisomic MI both had approx. 2.28 Xma per trivalent. In 4 out of 50 pentavalents (both Ts combined) the Xma occupied the same regional position in the trisomic portion of the pentavalent (between both Rb and the homologous acrocentric); in other words, the Xma tended to show regional clustering.

D. Analysis of MII. About half of all MII studied were fully analysable and from them we selected those that fulfilled certain criteria of 'completeness'. The accepted range was of cells with from 17 to 20 acrocentrics only, and with either one or two Rb added to the acrocentric counts. In all we accepted twelve different cell classes using orcein - a stain which precluded chromosome identification in the selected cells. Of the 79 analysable MII from Ts16, 48 fulfilled the criteria of complete cells as did 51 out of 65 Ts19 MII. The remaining 45 MII of the two trisomic systems were either so hypoploid as to be considered broken cells, or excessively hyperploid, presumably because arrested in anaphase I: these cells often contained both Rb. The distribution of cell types in the two Ts being similar, the pooled results of the 99 MII are set out in Table II.

TABLE II

MII observed and expected on two models of chromosome assortment (see text). Pooled cells of both trisomies. Complementary assortment types indicated by letters.

MII complements Rb + Acrocentric (Orcein-stained)	Numbers of MII Observed	Expected "Regular" assortment	Expected Random assortment
Rb + 19 (a)	11	20.34	18.6
2Rb + 17 (b)	14	19.80	3.1
Rb + 18 (a)	24	20.34	18.6
20 (b)	11	19.80	3.1
19 (c)	17	0.62	9.3
2Rb + 18 (c)	4	0.62	9.3
		"Irregular" assortment	
2Rb + 19 (d)	-	↑	9.3
18 (d)	4		9.3
2Rb + 20 (e)	1		3.1
17 (e)	4	18*	3.1
Rb + 20 (f)	1		6.2
Rb + 17 (f)	8	↓	6.2
Totals	**99**	**98.8**	**99.2**

* The good fit between this and the observed values reflects the assumption made about the expected numbers of irregular assortments.

Although the two Ts differ substantially in respect to the pairing behaviour at prophase within the Rb complex and in the parallel distribution of its chiasmate configurations at MI, the sorting out at anaphase I of chromosomes from the Rb complex, revealed at MII, may be assumed to follow the same pattern in the two Ts. Either assortment follows some standard "rules" of chromosome separation behaviour, or assortment of Rb and acrocentrics from the complex is entirely at random, or a mixture of the two leads to partly "regular" and partly "irregular", or even random, distributions. We shall contrast only the last two listed models. Table II shows the twelve orcein stained cell classes (out of a possible total of 32). It is assumed that the chromosomes of the standard seventeen acrocentric bivalents segregate regularly, then (v.s.) two sets of expected values were computed and contrasted with those observed. The first, broadly speaking "regular", is based on the simplistic "rule" that chromosomal elements linked by Xma will separate from each other at anaphase I. (The assortments are, of course, complementary, in pairs, and derived from the five types of configurations observed at MI, which are proportionally different in the two Ts, and taken into account in working out the expected values). In addition to the three pairs of complementary and strictly "regular" segregations, there are three pairs of "irregular" ones, for which no expectation could be computed, but they were assumed to represent one fifth of all assortments. The alternative set of expected values in Table II, given in the last column, is based on the assumption that chromosome assortment from the Rb complex is completely random for all cell types.

4. DISCUSSION

The EM study revealed different proportions of pentavalents and other pairing associations, and therefore, of univalence between the two Ts (Table I). The reasons for the differences are obscure, but might be related to timing of meiotic pairing which, apparently differs in the two Ts. Univalent axes have a general tendency to appear thick compared with the axes in bivalents or in triple associations; a phenomenon for which there is no ready explanation (see suggestions by Rasmussen et al., 1981; Speed, 1985). The trisomic axes of chromosomes 16 or 19, in the majority of pentavalents, are triply associated, and furthermore, the extent of triple association, though variable and somewhat different in the two Ts, is quite substantial. Because of the failure, with the method used, of demonstrating a central element, one might question whether two true and complete SC are running side by side in the regions of triple association. Inspection and measurement suggest that the axes in these regions are as close together as are the two axes in the bivalent region of pentavalents and of bivalents. Between these latter in the female mouse, two complete SC have been observed using other EM techniques (Nebel and Hackett, 1961). So our visual and measurement observations on the closeness of triple associations taken together with similar observations in the human trisomic female (Speed, 1984; Wallace and Hultén, 1983), as well as others (Holm and Rasmussen 1984), suggest that close triple pairing at meiosis does take place in many, if not all organisms when three homologues are present. It is likely, therefore, that, within the region of close triple pairing, true synapsis occurs and a complete SC forms. Clearly, and especially if a true SC were indeed demonstrated, the assertion that when three homologues face each other at meiotic pairing they exclusively pair in twos, would no longer be correct.

At MI the high frequency of pentavalents in Ts16, and the much lower one in Ts19, matches almost precisely the frequencies at prophase in the two Ts, so that once pairing has been established, crossing-over with Xma formation will ensue. Similarly, univalents and univalence at prophase persist into meiosis with their prophase frequencies and are seen at MI, with double the frequency in Ts19 than in Ts16 MI. By comparison, over sixty per cent of MI in free Ts21 in the human male show a univalent (Jagiello, 1981). One may expect, in general, an increase in Xma frequency in Ts, but the variation depends on species and on the specific chromosomes involved (Hewitt and John, 1965, Duck and James, 1976). In our material, Ts16 & Ts19 MI both show a slightly higher Xma frequency in the acrocentrics compared with balanced oocytes. In the only human female Ts21 cell reported, the Xma count was higher than average (Polani et al., 1982), while in human Ts21 males, the values little exceeded the normal (Hultén and Lindsten, 1970). In pentavalents, there is a tendency for Xma of the trisomic elements to be "clustered", and although the resolution of Xma position by microscopic analysis is imprecise, the fact that pairs of Xma in the trisomic element are assigned to the same homologous region with some frequency suggests that, contrary to what happens in regular bivalents with two Xma, there is less spacing out of Xma, so that Xma position interference (Whitehouse, 1973) may be less strict in triply paired chromosomes with SC lying side by side. In triploid Coprinus, (Rasmussen et al., 1981) recombination nodules in the parallel SC uniting the three axes of homologues, are distributed independently in each SC, suggesting that crossing over will not be subject to interference from one to the other SC.

MII express chromosome distribution from the MI, but together with others already mentioned, there are two main limitations to our analyses, i.e. our inability to utilise all the analyzable MII and, the model or models against which the results may be interpreted. We believe that cells discarded from analysis are random cells, so that no bias is introduced in the results. There remains the problem of how to interpret our results in terms of chromosome orientation and distribution. We have assumed that the chromosomes of the seventeen acrocentric bivalents disjoin regularly to the resulting MII, but for the Rb complex, it seems clear that the basic principles that guide orientation and distribution behaviour of chromosome multivalents at meiosis remain poorly understood. "Traditionally, explanation has been attempted by reference to certain .. features .. such as symmetry in chromosome size and centromeres position, chiasma frequency and location, and unspecified genotypic effects" (Rickards, 1983), to which centromere specificity, or, at any rate distribution-controlling functions may be added. Although when size differences of chromosomes are marked, symmetry differences have been interpreted to explain chromosome orientation and distribution oddities (Hamerton, 1970), in our material it is difficult to see how asymmetry may be invoked to explain the results of chromosome partitioning. We were, therefore, faced with a choice between two models with which to compare our observations: a model of complete randomness (see also Rickards, 1983); and a model of basically regular segregation from "rules" based on assortment from chiasmate configurations. While a completely regular chromosome assortment model would result in a deficit of some twenty per cent, and would be a very poor fit to the data, a model that admitted the more unbalanced, non-disjunctional assortments on an "irregular" basis, a mixed model, seemed more realistic. Expectations constructed from either of the two models are not a good fit to our observed distribution of orcein type MII (Table II), a not unusual problem in analyses of this type (Rickards, 1983). However, as one of the two models fits one part, and the other the other part of the data, we may tentatively suggest that a proportion of chromosome assortments may well be controlled by rules that determine chromosome orientation and subsequent separation; but that a not irrelevant proportion seems to assort irregularly, non-disjunctionally and, even possibly, at random. Some of the orientation and assortment problems that lead to the observed distribution of MII classes may derive from the very makeup of the Rb themselves, for example from their hypothetical centromere constitution that may be dual and exert variable influences on the mechanics of separation of paired chromosomes.

REFERENCES

Comings, D. E. and Okada, T. A. 1971. Triple chromosome pairing in triploid chickens. Nature 231, 119-121.

Duck, P. and James, A. P. 1976. The effect of aneuploidy on meiotic crossing over and segregation in yeast. Genet.Res. 28, 55-60.

Gropp, A. 1976. Morphological consequences of trisomy in mammals. In: Ciba Foundation Symposium, 40: Embryogenesis in Mammals, pp 155-175. Amsterdam: Elsevier/Excerpta-Medica/North Holland.

Gropp, A., Kolbus, U. and Giers, D. 1975. Systematic approach to the study of trisomy in the mouse II. Cytogenet.Cell Genet. 14, 42-62.

Hamerton, J. L. 1970. Robertsonian translocations. Evidence on segregation from family studies. In: Pfizer Medical Monographs No. 5: Human Population Cytogenetics (Jacobs, P. A., Price, W. H. and Law, P., Eds.) 63-80. Edinburgh University Press.

Hewitt, G. M. and John B. 1965. The influence of numerical and structural chromosome mutations on chiasma conditions. Heredity (Lond.) 20, 123-135.

Holm, P. B. and Rasmussen, S. 1984. The synaptonemal complex in chromosome pairing and disjunction. In: Chromosomes Today vol. 8, 104 - 116. (Bennett, M. D., Gropp, A. and Wolf, U., Eds.) London: George Allen & Unwin.

Hultén, M. and Lindsten, J. 1970. The behaviour of structural aberrations at male meiosis. Information from man. In: Pfizer Medical Monographs 5: Human Population Cytogenetics (Jacobs, P. A.,

Price, W. H. and Law, P. Eds.), 24 - 61. Edinburgh University Press.

Jagiello, G. 1981. Reproduction in Down syndrome. In: Trisomy 21 (Down syndrome) research perspectives. (de la Cruz, F. F. and Gerald, P. S. Eds.) 151-162. Baltimore: University Park Press.

Moses, M. J. and Poorman, P. A. 1984. Synapsis, synaptic adjustment and DNA synthesis in mouse oocytes. In: Chromosomes Today vol. 8, 90-103 (Bennett, M.D. Gropp, A. and Wolf, U. Eds.) London: George Allen & Unwin.

Moses, M. J., Poorman, P. A., Polani, P. E., Crolla, J. A., Moir, F., Davisson, M. T. and Roderick, T. H. 1982. Synapsis, synaptic adjustment and DNA synthesis in meiotic prophase of mouse oocytes. J.Cell Biol. 95, 77a.

Nebel, B. R. and Hackett, E. M. 1961. Synaptinemal complexes in primary spermatocytes of the mouse: the effect of elevated temperature and some observations on the structure of these complexes in control material. Z.Zellforsch. 55, 556-565.

Newton, W. C. F. and Darlington, C. D. 1929. Meiosis in polyploids. Part I. Triploid and pentaploid tulips. J.Genet. 21, 1-15.

Polani, P. E., Crolla, J. A., Seller, M. J. and Moir, F. 1979. Meiotic crossing over exchange in the female mouse visualised by BUdR substitution. Nature (Lond.) 278, 348 - 349.

Polani, P. E. and Jagiello, G. M. 1976. Chiasmata, meiotic univalents and age in relation to aneuploid imbalance in mice. Cytogenet.Cell Genet. 16, 505-529.

Polani, P. E., Dewhurst, J., Fergusson, I. and Kelberman, J. 1982. Meiotic chromosomes in a female with primary trisomic Down's syndrome. Hum.Genet. 62, 277-279.

Rasmussen, S. W., Holm, P. B., Lu, B. C., Zickler, D. and Sage, J. 1981. Synaptonemal complex formation and distribution of recombination nodules in pachytene trivalents of triploid Coprinus cinereus. Carlsberg Res.Commun. 46, 347-360.

Rickards, G. K. 1983. Orientation behaviour of chromosome multiples of interchange (reciprocal translocation) heterozygotes. Annu.Rev.Genet. 17, 443-498.

Speed, R. M. 1984. Meiotic configurations in female trisomy 21 foetuses. Hum.Genet. 66, 176-180.

Speed, R. M. 1985. The prophase stages in human fetal oocytes studied by light and electron microscopy. Hum.Genet. 69, 69-75.

Swanson, C. P., Merz, T. and Young, W. J. 1981. Cytogenetics, p 140, Englewood Cliffs, N.J., Prentice-Hall.

Wallace, B. M. W. and Hultén, M. A. 1983. Triple chromosome synapsis in oocytes from a human foetus with trisomy 21. Ann.Hum.Genet. 47, 271-276.

Whitehouse, H. L. K. 1973. Towards an understanding of the mechanism of heredity. pp114-115. London: E. Arnold Limited.

Meiotic chromosome pairing in the human male: experience from surface spread synaptonemal complexes

M. Hulten,[1] N. Saadallah[1,2] and J. Batanian[1,2]

[1]*Regional Cytogenetic Laboratory, East Birmingham Hospital, Birmingham B9 5ST, England, and* [2]*Genetics Department, University of Birmingham, P.O. Box 363, Birmingham B15 2TT, England*

The surface spreading technique has allowed a more detailed account of the pairing of chromosomes at first meiotic prophase than has been possible by traditional investigations (review in Moses et al 1985) and gives complementary information to that obtained by the more time consuming serial sectioning (review in Rasmussen and Holm 1980). This presentation will give a broad outline of the most interesting findings on human male spermatocytes using modifications of the surface spreading techniques by Moses (1977), Fletcher (1979) and Solari (1980) as detailed in Hulten et al (1986). The material investigated is given in Tables 1 and 2. Most men were ascertained for a disturbance in reproduction and further investigations were undertaken when a constitutional chromosomal rearrangement was revealed. As it is not easy to obtain testicular biopsies from an entirely adequate normal material of fertile men, 'controls' were men with an apparently normal mitotic karyotype and normal testicular histology without obvious signs of impaired spermatogenesis. Most investigations concentrated on the pachytene pairing stage, and there is less information on the previous leptotene and zygotene or the following diplotene stages.

ALIGNMENT AND HOMOSYNAPSIS

Homologues are aligned by their attachment to the nuclear membrane at both ends, and presumably this is the main reason why homologous synapsis (homosynapsis) generally starts at the termini proceeding towards the centre of the bivalents with the pericentromeric regions being the last to pair. The regular occurence of trivalents in Robertsonian, quadrivalents in simple and especially the hexavalents in one complex three breakpoint reciprocal translocation as illustrated in Fig. 1 indicates that homosynapsis is highly efficient. Investigations in trisomy 21 demonstrate in addition that, contrary to the conventional view, all three homologues may be kept in register.

The behaviour of inversions give further information on interstitial pairing initiation (Fig. 2). This is comparatively rare in control spermatocytes but is regularly seen in the three loop forming inversions so far investigated (Table 2). The simplest interpretation of the configurations of eight in Fig. 2a and b is that they result from alignment and synapsis of homologous sequences in the middle of the inverted segment along their entire length. This demonstrates a remarkable exactness and fidelity of homologous sequences in attracting each other, especially as a foldback is required in the middle of the inverted segment for homosynapsis to take place. The same behaviour has been seen in the three different loop forming pericentric inversions, and there is thus no indication of specialised interstitial pairing segments along the length of the chromosome (see e.g. Maguire, 1986). The occurence of asymmetric loop configurations such as that in Fig. 2c made Gabriel-Robez et al (1986) propose a polarity in homosynapsis, being more rapid in one direction. A catch-up may however take place before pairing in the leading direction has been completed, as exemplified by the symmetrical configuration in Fig. 2d.

MECHANICAL CONSTRAINTS, CONSTITUTIVE HETEROCHROMATIN AND ASYNAPSIS

Trivalents in Robertsonian translocations may be fully homosynapsed, but the central homologous segments of the quadrivalents of simple reciprocal translocations and in particular of the hexavalents in the complex translocation often remain unpaired (Fig. 1). Presumably this is dependant on the mechanical constraint in these more complicated pairing situations. The pairing behaviour of the pericentric inversions (Table 2) reinforces this interpretation as loops were rarely found in the smallest occupying only about one per cent of the haploid autosomal genome, while loop formation was regular in all three larger inversions, sized five to six per cent.

Heterochromatic blocks may seemingly contribute to and enhance these mechanical problems. In the normal situation the 1qh, 9qh, 16qh and the p arms of the acrocentric chromosomes 13-15 and 21-22 are often delayed in synapsis, and the same holds true when they are involved in a structural rearrangement. These segments may then remain asynaptic and the proper homologous synapsis of adjacent euchromatic regions sometimes be prohibited (Fig. 2 d,g).

Autosomal segments remaining unsynapsed are easily recognised by their typical morphology, showing so called excrescences, granulae of loops similar to those of the differential axes of the XY bivalent (compare Fig.1f,g and Fig. 2 a,c,d,g).

SECONDARY HETEROSYNAPSIS

Following the initial homosynaptic stage at early pachytene, lateral elements remaining unsynapsed may pair with any available partner including the non-homologous (heterosynapsis). In the process the excrescences fuse and eventually disappear. The hetero-synapsed Synaptonemal Complex so created is indistinguishable from the 'normal' homosynapsed (Fig. 2h). Judging from the behaviour of the inv(13) this secondary non-homologous zipping-up procedure is rapid, once permitted, as there are no intermediate configurations between the forked, where there is an apparent blockage to heterosynapsis (Fig. 2g), and the rods with complete heterosynapsis (Fig. 2h).

INTERCHROMOSOMAL EFFECTS AND SPERMATOCYTE SURVIVAL

One of the structural heterzygotes, the inv(1)(p32q42) showed not only a lack of synapsis within the inversion loop but also a general disturbance in pairing with both asynapsis and precocious desynapsis of other homologues. We have tentatively classified this as an interchromosomal effect secondary to the primary failure of synapsis of the inverted segment, and suggested that this may contribute to the maturation arrest of spermatogenesis seen in this case.

Both the inv(2) and the inv(13) showed some irregularity in pachytene timing, and some pertuberance of the meiotic clock is likely to be a general feature of constitutional chromosomal aberrations. Nevertheless they were both fertile, and in contrast to the infertile inv(1) heterozygote, generally characterised by heterosynapsis at later stages of pachytene. On the basis of this discrepancy we may conclude that heterosynapsis may play a functional role in rescuing spermatocytes from the potential damage of asynapsis, and allow maturation to proceed. The price to be paid for the retained spermatogenesis might be increased non-disjunction, if we consider heterosynapsis analogous to distributive pairing (Grell 1971,1985).

The association between structurally rearranged chromosomes and the XY may be a special case of heterosynapsis (Fig.1 d,e) without any particular relevance for spermatocyte survival (Rosenmann et al 1985). In our own material there is no obvious correlation between such XY association and impairment of spermatogenesis (Table 3).

RECOMBINATION

Lack of synapsis such as in the small inv(13) is expected to lead to cross-over supression and absence of recombinant offspring. However, in this particular case three out of eight pregnancies from two male carriers were recombinant (unpubl. obs.), and we have been wondering if they might be the result of a U-type exchange rather than a traditional cross-over. A detailed investigation on the frequency distribution of so called bars or recombination nodules at pachytene in relation to chiasmata at diakinesis/first metaphase would have been valuable but has yet not been performed, neither in structural heterozygotes nor in the normal situation. The preliminary impression is however, that there is a reasonable agreement between these two parameters (Solari, 1980;Saadallah and Hulten, 1986), as would be expected if they both identify sites of crossing-over.

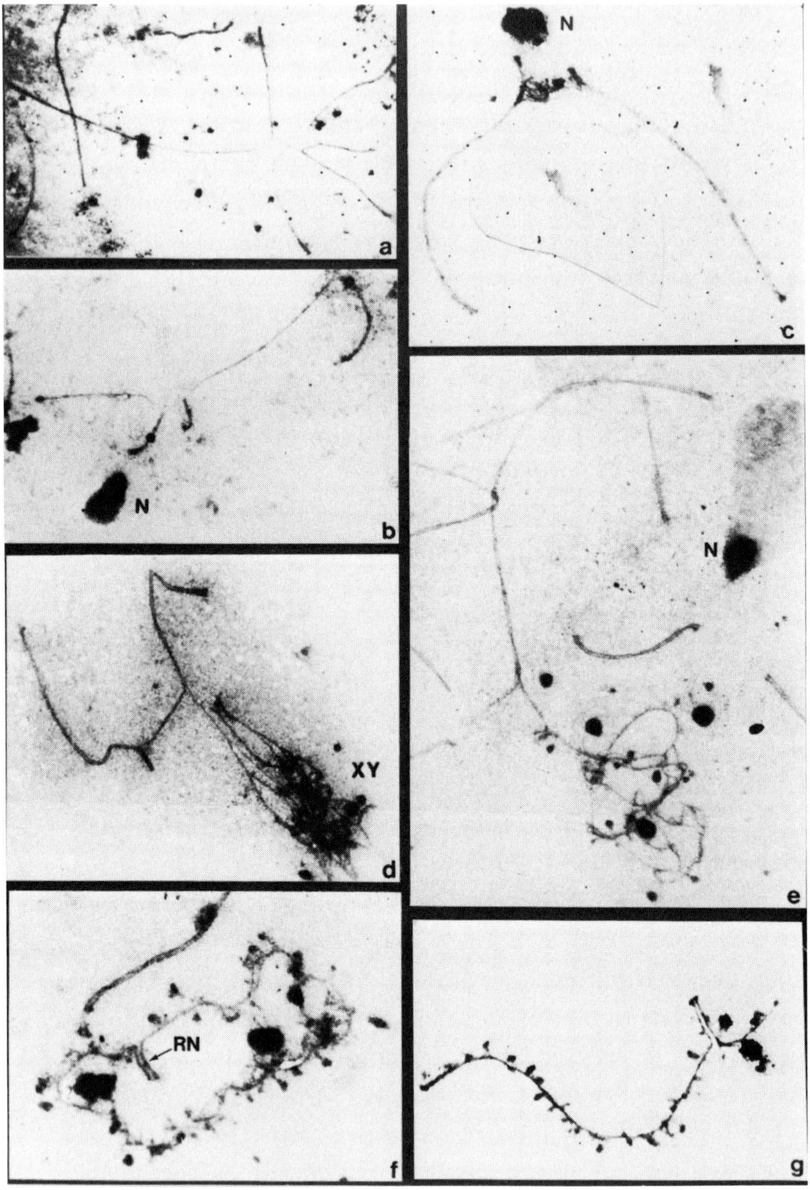

Fig. 1 (left). EM synaptonemal complexes. (a) Hexavalent from 46,XY,t(2;4;9);(b) Quadrivalent from 46,XY,t(9;21);(c and d) Trivalents from 45,XY,t(13;14) with central part unpaired in (c) and association to XY at late pachytene in (d);(e) apparent heterosynapsis between autosomal bivalent and XY at late pachytene; (f and g) XY bivalents at mid (f) and early pachytene (g) both showing pairing segment and a possible recombination nodule in (f).

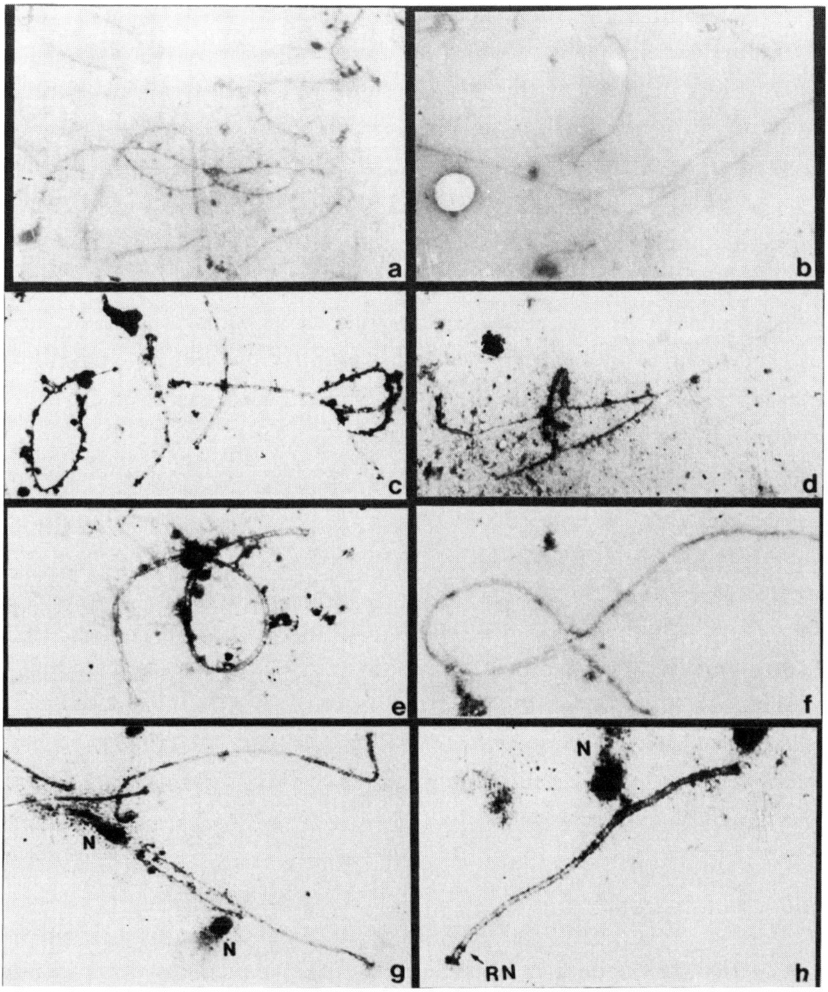

Fig. 2 (above). EM inversion bivalents showing synaptic initiation (a,b), progression (c,d) and loops (e,f), total asynapsis (g) and full heterosynapsis (h).
Examples in a,d,e are from the inv(1)(p32q42), in b,c and f from the inv(2)(p13q35) and g,h from the inv(13)(p12q14).
N = nucleolus organiser. RN = recombination nodules.

COMMENT AND SUMMARY

This short review has concentrated on first meiotic synapsis in the human male and the new information obtained by investigation of surface spread spermatocytes. The same technology may with slight modifications be used in the female, fetal oocytes(ref. in Hulten et al 1986). Synapsis in general proceeds in the same way in both sexes but interstitial initiation seems to be slightly more common in oocytes. The pachytene synaptonemal complexes in oocytes are also longer, possibly thereby accommodating a higher amount of recombination. Fetal trisomy 21 has been analysed in some detail, demonstrating that all three homologues may regularly be held in registry, sometimes along their entire length. As yet no structural heterozygotes have been investigated, and the access to such material is likely to be limited.

In summary, by application of the surface spreading which is a comparatively rapid and relatively straight forward technique, we have learnt a great deal more about alignment and homologous recognition, and about the conditions for homo and heterosynapsis. The mechanisms behind are however still not well understood, and remain a challenge. Neither do we have any detailed knowledge on chiasma formation or the effect of pertuberations of synapsis for first meiotic chromosome segregation and non-disjunction, which would be particularly important in view of the high amount of aneuploidy in the human.

ACKNOWLEDGMENTS

This work was supported by student scholarships from the Lebanese Government to N.S and J.B. The present address of N.S is Lebanese University, Sciences Faculty II, PO Box: 33, Mansonneh ELMETEN, Lebanon.

REFERENCES

Batanian, J. and M.A. Hulten 1986. EM investigations of synaptonemal complexes in an infertile human male carrier of a pericentric inversion inv(1)(p32q42). Submitted to Hum. Genet.

Chandley, A.C., P. Goetz, T.B. Hargreave, A.M. Joseph and R.M. Speed 1984. On the nature and extent of XY pairing at meiotic prophase in man. Cytogenet. Cell Genet. 38, 241-247.

Chandley, A.C., R.M. Speed, S. McBeath and T.B. Hargreave 1986. A human 9;20 reciprocal translocation associated with male infertility analysed at prophase and metaphase I of meiosis. Cytogenet. Cell. Genet. 41, 145-153.

Fletcher, J.M. 1979. Light microscope analysis of meiotic prophase chromosomes by silver staining. Chromosoma (Berl) 72, 241-248.

Gabriel-Robez, O., C. Ratomponirina, Y. Rumpler, B. Le Marec, J.M. Luciani, M.R. Guichaoua 1986. Synapsis and synaptic adjustment in an infertile human male heterozygous for a pericentric inversion in chromosome 1. Hum. Genet. 72, 148-152.

Grell, R.F. 1971. Distributive pairing in man? Ann. Genet. 14, 165-171.

Grell, R.F. 1985. The meiotic process and aneuploidy. In Aneuploidy etiology and mechanisms, V.L. Dellarco, P.E. Voytek and A. Hollaender, ed., Basic life sciences 36, Plenum Press, (New York and London) pp 317-335.

Hulten, M.A., N. Saadallah, B.M.N. Wallace and M.R. Creasy 1986. Meiotic studies in man. In Human Cytogenetics, a practical approach. D.E. Rooney and B.H. Czepulkowski, ed. IRL press, (Oxford, Washington DC) pp 163-196.

Johannisson, R., A. Gropp, H. Winking, W. Coerdt, H. Rehder, and E. Schwinger 1983. Down's Syndrome in the male. Reproductive pathology and meiotic studies. Hum. Genet. 63, 132-138.

Laurent, C., A.C. Chandley, B. Dutrillaux and R.M. Speed 1982. The use of surface spreading in the pachytene analysis of a human t(Y;17) reciprocal translocation. Cytogenet. Cell Genet. 33, 312-318.

Maguire, M.P. 1986. The pattern of pairing that is effective for crossing-over in complex B-A chromosome rearrangements in maize III. Possible evidence for pairing centers. Chromosoma 94, 71-85.

Moses, M.J. 1977. Synaptonemal complex karyotyping in spermatocytes of the Chinese hamster. Chromosoma 60, 99-125.

Moses, M.J., P.A. Pourman, M E. Dresser, G.K. De Weese and J.B. Gibson 1985. The synaptonemal complex in meiosis: significance of induced pertuberance. In: Aneuploidy, etiology and mechanisms. V.L. Dellarco, P.E. Voytek and A. Hollaender, ed. Basic life sciences, 36, Plenum Press, (New York and London) pp 337-352.

Rasmussen, W. and P.B. Holm 1980. Mechanics of meiosis. Hereditas 93, 187-216.

Ratomponirina, C., J. Couturier, O. Gabriel-Robez, Y. Rumpler, B. Dutrillaux, M. Croquette, Q. Rabache and M. Leduc 1985. Aberrations of the synaptonemal complexes in male 46,XY,-14,+der (14)t(Y;14) . Ann. Genet. 28, 214-218.

Rosenmann, A., J. Wahrman, C. Richler, R. Voss, A. Pensitz and B. Goldman 1985. Meiotic association between the XY chromosomes

and unpaired autosomal elements as a cause of human male sterility. Cytogenet. Cell Genet. 39, 19–29.
Saadallah, N. 1985. PhD thesis, University of Birmingham.
Saadallah, N. and M.A. Hulten 1985. A complex three breakpoint translocation involving chromosomes 2,4 and 9 identified by meiotic investigations of a human male ascertained for subfertility. Hum. Genet. 71, 312–320.
Saadallah, N. and M.A. Hulten 1986. EM investigations of surface spread synaptonemal complexes in a human male carrier of a pericentric inversion inv(13)(p12q14). The role of heterosynapsis for spermatocyte survival. Ann. Hum. Genet. 50, 369–383.
Solari, A.J. 1980. Synaptonemal Complexes and associated structures in microspread human spermatocytes. Chromosoma (Berl) 81, 315–337.
Templado, C., F. Vidal, J. Navarro, S. Marina and J. Egozcue 1984. Meiotic studies and synaptonemal complex analysis in two infertile males with a 13/14 balanced translocation. Hum. Genet. 67, 162–165.
Vidal, F., J. Navarro, C. Templado, S. Brusadin and J. Egozcue 1984. Synaptonemal complex studies in a mosaic 46,XY/47,XXY male. Hum. Genet. 66, 306–308.
Vidal, F., C. Templado and J. Navarro 1982. Meiotic and synaptonemal complex studies in a 14/21 translocation carrier. Int. J. Androl. 5, 21–26.

Table 1 Material of surface spread spermatocytes

	No of	Karyotype	Means of analysis	No. of nucl.	Reference
Controls	1	46,XY	EM	22 [86]	Solari (1980)
	26	46,XY	EM	20 [891]	Chandley et al (1984)
	8	46,XY	LM+EM	323 [16]	Saadallah (1985)
Numerical	1	47,XY,21+	EM	27 [55]	Johanisson et al (1984)
	1	46,XY/47,XXY	LM+EM	32	Vidal et al (1984)
Rob. transl.	1	45,XY,t(14q21q)	LM	99	Vidal et al (1982)
	2	45,XY,t(13q14q)	LM+EM	100	Templado et al (1984)
	2	45,XY,t(14q21q)	LM+EM	210	Rosenmann et al (1985)
	2	45,XY,t(13q14q)	EM	34 [8]	Batanian and Hulten (in prep.)
Rec. transl.	1	46,XY,t(2;4;9)(p13;q45;p12)	LM+EM	163 [78]	Saadallah and Hulten (1985)
	1	46,XY,t(4;16)(p14;p11.2)	EM	9 [1]	Batanian and Hulten (in prep.)
	1	46,XY,t(9;20)(q34;q11)	EM	100	Chandley et al (1986)
	1	46,XY,t(9;21)(q22.33;q11.2)	EM	9 [10]	Batanian and Hulten (in prep.)
	1	46,X,t(Y;17)(q11.21;q12)	LM	15	Laurent et al (1982)
	1	46,XY,der(14),t(Y;14)(q11;p11)pat	EM	3	Ratonponirima et al (1985)

LM = light microscopy, EM = electron microscopy. Numbers within brackets [ ] indicate additional incomplete nuclei assessed only regarding XY morphology.

Table 2 EM analyses of pericentric inversions: Synapsis of inverted segment

Karyotype	No of nucl.	Homosynapsis: Asyn. (%)	Homosynapsis: Init. (%)	Homosynapsis: Progr. (%)	Homosynapsis: Loop (%)	Heterosynapsis: Small Loop (%)	Heterosynapsis: Rod (%)	Reference
46,XY,inv(1) (p32q12)(16qh+	30	0 (0)	11(37)	13(43)	1(43)		5 (17)	Gabriel-Robez et al (1986)
46,XY,inv(1) (p32q42)	134[84]	18(13)	28(21)	68(51)	68(51)	0 (0)	0 (0)	Batanian and Hulten (1986)
46,XY,inv(2) (p13q35)	38[20]	0 (0)	6(16)	12(37)	7(18)	6(16)	7(18)	Hulten and Saadallah(in prep.)
46,XY,inv(13) (p12q14)	97[76]	38(39)	3 (3)	0 (0)	1(1)	0 (0)	55(57)	Saadallah and Hulten (1986)

Numbers within brackets [ ] indicate additional incomplete nuclei assessed only regarding XY morphology.

Table 3

Ascertainment	Karyotype	Spermatogenesis	XY association			
			Struct. rearr. (%)	Unspec. (%)	Non-ass. (%)	Total
Controls: infert(4) test op. for other reasons(4)	46,XY	no obv. abn.	–	63 (19)	276 (81)	339
Child with NTD	46,XY,(inv2)(p13q35)	some degen.	3 (5)	29 (50)	36 (45)	58
Prev. child with chrom. abn.	(46,XY,inv(13)(p12q14)	slight degen.	19 (11)	36 (21)	118 (68)	173
	(46,XY,t(9q;21q)	some degen.	2 (11)	6 (32)	11 (58)	19
Habitual abortion	(46,XY,t(13q14q)	oligo-normospermia	7 (27)	11 (42)	8 (31)	26
	(46,XY,t(4p;16p)	normospermia	1 (10)	5 (50)	4 (40)	10
	(46,XY,t(2p;4q;9p)	normospermia	1 (1)	31 (22)	112 (78)	144
Infertility	(46,XY,t(10q17q)	mat. arrest	5 (31)	6 (38)	5 (31)	16
	(46,XY,(inv1)(p32q42)	first spermat.	1 (1)	12 (13)	81 (86)	94

The oocyte spindle and predisposition to aneuploidy in mice

U. Eichenlaub-Ritter

Universität Bielefeld, Biologie XI, Gentechnologie, P.O. Box 8640, 4800 Bielefeld 1, Federal Republic of Germany

Several hyothesis have associated the maternal-age related rise in aneuploidie in mammals with events taking place during different stages of oocyte development (for reviews see: Bond and Chandley, 1983; Hassold and Jacobs, 1984). For instance, the production line hypothesis proposes that a decrease in cross-over activity during early stages in meiosis in those oocytes in the embryo determined to become ovulated late in life increases the number of univalents. This is supposed to influence the equal distribution of chromosomes. Other theories relate induction of aneuploidy to damage of the oocyte due to environmental hazzards during its long resting and growth in the ovary before ovulation. Hormonal imbalance in organisms approaching the end of their reproductive lifespan is also thought to participate in aetiology of aneuploidy. It seems to affect mainly maturation.

However, one has to keep in mind that regular distribution of chromosomes into two daughter cells is ultimately facilitated by the spindle apparatus and cytoskeleton of the oocyte. Therefore, in the end, some alteration in its structure and/or components must be involved in induction of aneuploidy. In fact, it has been demonstrated that predominantly late stages in the formation of the female gamete seem to be susceptible to exposure to drugs, particularly such interferring with the polymerization of tubulin (Sugawara and Mikamo, 1980; Kaufmann, 1985). However, no one has so far examined the spindle apparatus itself with regard to maternal age and induction of aneuploidy. Employing immunofluorescence techniques we were able to detect alterations of the microtubular cytoskeleton and chromosome alignment of spontaneously ovulated oocytes due to postovulatory ageing, and in relation to maternal age in CBA/Ca mice. These animals were taken as a model system since an increase in aneuploidy in relation to maternal age but also physiological ageing and depletion of the ovary has been reported in this strain of mice (Brook et al., 1984).

SPINDLE STRUCTURE AND CHROMOSOME ALIGNMENT OF FRESHLY OVULATED AND POSTOVULATORY AGED OOCYTES

Although age-related induction of aneuploidy seems to be mainly based on errors in chromosome distribution at the first maturation division (reviewed in Hassold and Jacobs, 1984), we examined initially naturally ovulated oocytes of young mice blocked in metaphase II. Ageing of these eggs in the fallopian tubes is also enhancing the incidence of abnormal development of the embryo and probably aneuploidy although it affects the cell irrespective of maternal age (Gianfortoni and Gulyas, 1985). We compared three groups of experimental cells. First, we looked at oocytes which had been ovulated recently, isolated from the ampullae of young mice early on the day of oestrus. Secondly, intermediately aged oocytes starting to loose their cumulus were taken from mice late in the day of oestrus. Thirdly, we examined oocytes which were freed from the fallopian tubes of animals on the day after oestrus, thus aged for more than 24 hrs.

Indirect tubulin immunofluorescence shows that microtubules are nearly exclusively associated with the barrel-shaped spindle in freshly ovulated oocytes.

Only faint staining of several focal areas in the cytoplasm indicate that some tubulin becomes assembled also at cytoplasmic centrosomes in these cells. Fluorescence between polar centrosomes and chromosomes in the spindle is very bright and evenly distributed. No astral microtubules are present in spindles of recently ovulated oocytes which run from the spindle poles into the cytoplasm, and away from the chromosomes.

In intermediately aged cells which start to loose their cumulus, the spatially controlled polymerization of microtubules between chromosomes and polar centrosomes seems to be gradually lost in favour of uncontrolled assembly of tubules at organizing sites all over the cell. This leads to the formation of prominent cytoplasmic asters with long microtubules while tubules in the spindle appear to be lost. Astral fibers form which run from the spindle poles into the cytoplasm and intermingle with other cytoplasmic tubules. In

Fig. 1: Indirect tubulin immunofluorescence of spindles in freshly ovulated mouse oocytes: Spindle poles are usually flat (1a) but may also be slightly more fusiform (1c). Chromosomes are all aligned at the spindle equator (1b; double exposure after FITC/antitubulin, Dapi staining).
Fig. 2: The cytoplasm of freshly ovulated oocytes is nearly devoid of tubules. Only faintly staining foci recognized by antibodies against tyrosylated (2b) and detyrosylated (2c) α-tubulin are apparent. (2a: Dapi stained chromosomes in the same cell).
Fig. 3 and 4: Cytoplasmic asters form in oocytes aged for several hours in the fallopian tubes (fig. 3) but their number decreases upon prolonged ageing (more than 24hrs, fig. 4).
Fig. 5: Postovulatory ageing induces loss of microtubules, formation of aster microtubules in the cytoplasm (5a) and at the spindle poles (5a, b), and gradual degeneration of the entire spindle (5d, f). In a high proportion of oocytes chromosomes are no more located centrally between the poles, at the equator (5c, d, g). (b, c; d, e; f, g are corresponding images FITC/tubulin immunofluorescence and Dapi of the same cell).
Bar in all figures: 10µm.

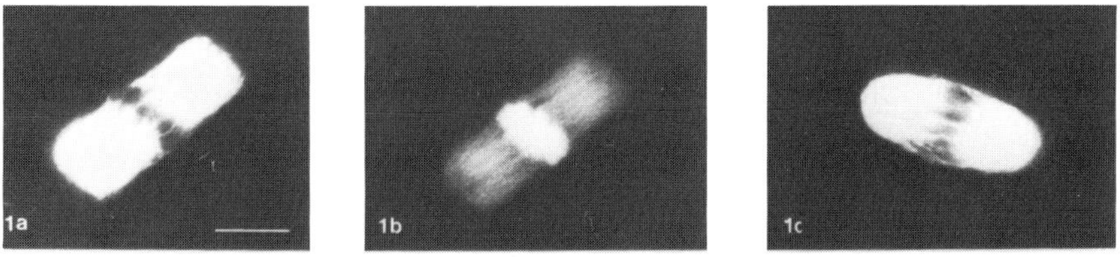
1a
1b
1c

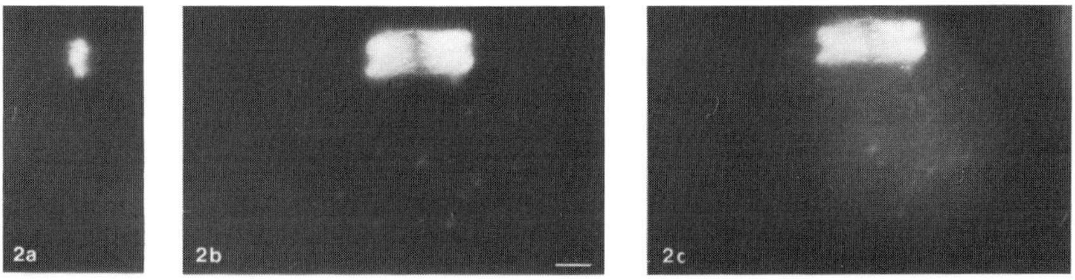
2a
2b
2c

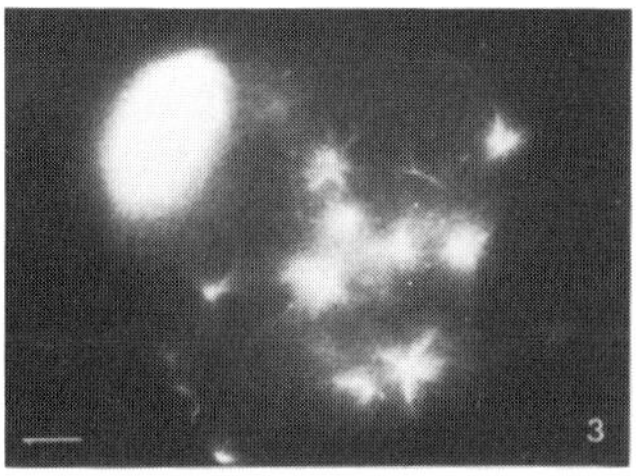
3

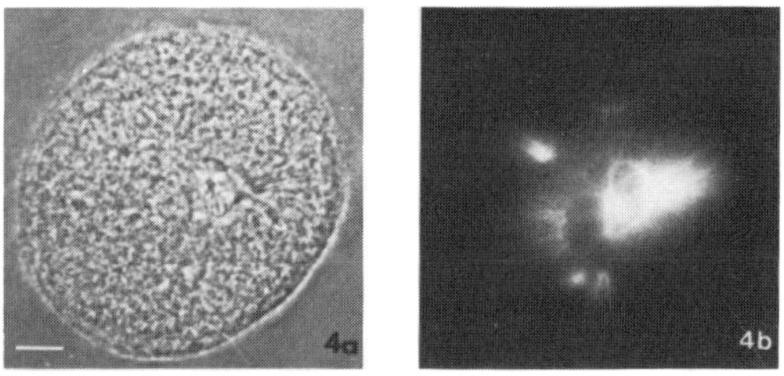
4a
4b

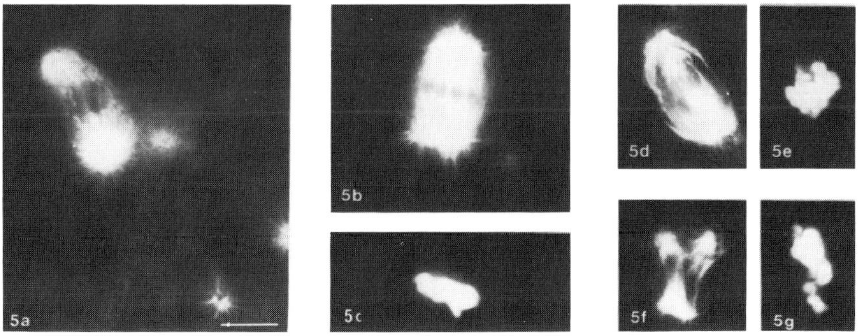
5a
5b
5c
5d
5e
5f
5g

the spindle fluorescence is predominantly lost in the vicinity of the chromosomes around the equator.

In postovulatory aged cells which were free of cumulus when taken from the fallopian tubes spindles become generally more fusiform and smaller. Cytoplasmic asters may be present in some cells but their number appears reduced. Others apparently lose all of them. Measurements of the pole-to-pole distance of spindles in freshly ovulated oocytes reveal that the spindle becomes shortened during postovulatory ageing. The average spindle length of 26,4µm in freshly ovulated cells decreases significantly to 21,7µm in postovulatory aged oocytes (t-test: $p < 0{,}001$).

Studies using antibodies against tyrosylated and detyrosylated α-tubulin imply that the alterations in the oocytes microtubular cytoskeleton do not seem to depend on a posttranslational modification of α-tubulin at its carboxy-terminus but may more likely be caused by different activities of microtubule organizing centers.

Chromosomes in freshly ovulated cells are nearly always properly aligned at the spindle equator. In only 11 % of all cells we observed, one or several chromosomes were not located precisely at the center of the spindle. In aged oocytes, however, disorder becomes greatly increased. In 61,4 % of all spindle chromosomes were dislocated from the spindle equator.

These observations demonstrate how natural ageing may affect the spindle. It is concluded that ageing probably impairs the ability of chromosomes, or rather kinetochores, to direct alignment of and provide attachment for spindle fibers. This could, in turn, predispose to nondisjunction.

SPINDLES AND CHROMOSOMES IN SPONTANEOUSLY OVULATED OOCYTES OF YOUNG, OLD AND UNILATERALLY OVAIOECTOMIZED CBA MICE

The overall morphology of metaphase II spindles is similar in oocytes from young and old mice. As characteristic for recently ovulated oocytes they all do not possess conspicious asters in their cytoplasm or astral fibres coming off the spindle poles. However, measurements of spindle length from 1 1/2 to 4 month old mice and such that were 9 month or older revealed that spindles in the oocytes from old animals are significantly shorter than those of young mice (22,7μm versus 26,4μm; t-test: $p < 0,001$) (Eichenlaub-Ritter et al., in preparation). Similarly, the pole-to pole distance of spindles in oocytes derived from 6-7 month old mice which had been unilaterally ovarioectomized and approached the end of their reproductive life was also less than in the young animals (23,3μm versus 26,4μm; t-test: $p < 0,001$). Spindle length may be related to age since oocytes from 6-7 month old sham controls which had been only injected with anaethetics like the oviex. mice but otherwise untreated exhibit also shorter spindles than young mice (23,0μm versus 26,4μm; t-test: $p < 0,001$; no significant difference between 6-7 month old ovarioectomized and sham operated mice). In contrast, a feature which is not correlated with age but with nearness to the end of the reproductive lifespan and also to predisposition to aneuploidy (Brook et al., 1984) is the degree of order of chromosome alignment on the spindle in the oocyte. Thus, while 11 % of oocytes of young mice (up to 4 month)show irregularities in alignment, and this is also observed in only 22,7 % of the 6-7 month old sham controls, there is a sharp rise in percentage of spindles with displaced chromosomes in oocytes of 9 month old CBA mice and the 6-7 month old unilaterally ovarioectomized animals (64,9 % and 58,3 %, respectively).

OUTLOOK

Our data show that in vivo ageing of mouse oocytes is accompanied by distinct alterations in the spindle apparatus of the oocyte. The most prominent features of structural changes observed in the aged oocytes are I) a loss of microtubules in the vicinity of the chromosomes, in the center of the spindle, II) formation of astral fibres at the poles, and III) appearance of large cytoplasmic asters (Eichenlaub-Ritter et al., submitted). This seems to reflect the gradual loss of activity of kinetochores and/or chromosomes to direct microtubule outgrowth and positioning. It is suggested that dislocation of microtubules from the spindle into the cytoplasm in moderately aged oocytes and less favourable conditions for assembly of tubulin in even longer postovulatory aged oocytes enhances chances for nondisjunction and malsegregation of chromosomes and predisposes to aneuploidy. The degree of chromosome misalignment is under these conditions indicative of the risk for the chromosomes to become malsegregated at next division.

Since most errors in chromosome distributions leading to material age-related aneuploidy seem to occur at the first meiotic maturation division, at least in man, (Hassold and Jacobs, 1984) it was not expected to find structural differences in spindles of oocytes from young and old mice which are formed only during the second meiotic division. However, the observed decrease in spindle length in correlation with maternal age indicates that there may be some parameters which influence oocyte development and become changed due to ageing of the female, irrespective of the nearness to the end of the reproductive life and physiological age of the ovary. This does not seem to expose oocytes per se to aneuploidy. As earlier studies have shown, incidence of aneuploid offsping in untreated 6-7 month old CBA mice is still quite low, although they have shorter spindles than young mice, and as long ones as the oviex. mice with many more chromosomally unballanced offspring (Brook et al., 1984). Shorter spindles and, possibly less favourable conditions for tubulin assembly could, however, make oocytes of older females more sensitive to environmental interference, which, in turn may cause aneuploidy. In fact, altered sensitivity to colchicine with age has been demonstrated in other strains of mice (Tease and Fisher, 1986).

As a more direct indicator of predisposition to aneuploidy, one can take the order of chromosome alignment in oocytes. In accordiance with recent findings that imply that physiological age of the ovary determines predisposition to aneuploidy (Brook et al., 1984), we observe a direct correlation between occurance of chromosome displacement in metaphase II spindles and depletion of the ovary/nearness to the end of the reproductive lifespan. In contrast to the loss of order in chromosome alignment of postovulatory aged cells, displacement of chromosomes in fresh oocytes of old animals is probably related to a preceeding non-disjunction event. We have meanwhile started to look at maturing oocytes in meiosis I. We hope to be able to detect structural differences which enable us to trace the cause for malsegregation of chromosomes and get some indication for the factors responsible for the maternal age-related increase in aneuploidy. Up to now preliminary data indicate that the dynamics of different processes taking place after reinitiation of meiosis and during maturation of the oocyte, as there are germinal vesicle breakdown, spindle formation and polar body extrusion,are altered in oocytes of old females. Therefore one may only speculate that, in analogy with the structural changes found in postovulatory aged oocytes, the stringent spatio-temporal control of tubulin polymerization exerted by the chromosomes, kinetochores and centrosomes could be disturbed in oocytes of females approaching the end of their reproductive life, thus predisposing the cell to aneuploidy.

Acknowledgements: I am most grateful to Drs. A.C. Chandley and R.G. Gosden for their interest, discussions and advice throughout the course of the work. The anti-tubulin antibodies were kind gifts of Drs. B. Jockusch, J. Kilmartin and J.C. Bulinski. The work has been supported by a fellowship of the German Research Foundation.

REFERENCES:

Bond, D.J., Chandley, A.C.: Aneuploidy. Oxford Univ. Press, Oxford 1983.

Brook, J.D., Gosden, R.G., Chandley, A.C., 1984: Maternal ageing and aneuploid embryos - Evidence from the mouse that biological and not chronological age is the important influence. Hum.Genet. 66: 41-45.

Eichenlaub-Ritter, U., Chandley, A.C., Gosden, R.G.: Alterations of the microtubular cytoskeleton and increased disorder of chromosomal alignment in spontaneously ovulated mouse oocytes aged in vivo: an immunofluorescent study. Chromosoma (Berl.) submitted.

Eichenlaub-Ritter, U., Chandley, A.C., Gosden, R.G.: Chromosome alignment and spindle structure in spontaneously ovulated oocytes of young, old and unilaterally ovarioectomized mice: spindle length and chromosome dislocation as indicators of predisposition to aneuploidy. Chromosoma, in preparation.

Gianfortoni, J.G., Gulyas, B.J., 1985: The effects of short term incubation (ageing) of mouse oocytes on in vitro fertilization, zona solubility, and embryonic development. Gamete Res. 11: 59-68.

Hassold, T.J., Jacobs, P.A., 1984: Trisomy in man. Ann.Rev.Genet. 18: 69-97.

Kaufmann, N.H., 1985: An hypothesis regarding the origin of aneuploidy in man: indirect evidence from an experimental model. J.Med.Genet. 22: 171-178.

Sugawara, S., Mikamo, K., 1980: An experimental approach to the analysis of mechanisms of meiotic nondisjunction and anaphase lagging in primary oocytes. Cytogenet. Cell Genet. 28: 251-264.

Tease, C., Fisher, G., 1986: Oocytes from young and old mice respond differently to colchicine. Mutation Res. 173: 31-34.

NUCLEOLUS ORGANISER

The molecular organization of the human ribosomal gene

R. D. Schmickel

Department of Human Genetics, University of Pennsylvania, School of Medicine/G3, 37th and Hamilton Walk, Philadelphia, PA 19104, USA

One of the attractive features of ribosomal genes is that they can be analyzed at both a cytological and a molecular level. Cytologically the ribosomal genes can be demonstrated by in situ hybridization. In metaphase the appear as the 1-5 micron stalks on the short arm of the acrocentric chromosomes. In interphase cells they appear as fibrillar centers measuring 0.5 to 5 microns in diameter (Stahl 1980). The association of the acrocentric chromosomes in nucleoli and at metaphase is well known. There is also a tendency for chromosomal recombination between the short arms of the acrocentric chromosomes which often results in the loss of ribosomal genes. The ribosomal genes are located on the short arm of the acrocentric chromosomes between the distal fluorescent chromatin and the centromeric alphoid heterochromatin. The appearance of this region is highly variable and serves as a marker for nondisjunction (Magenis and Chamberlin 1981).

Each of these cytogenetic characteristics can be described at a molecular level. The ribosomal genes are located on each of the acrocentric chromosomes (Henderson et al. 1972). Quantitation has shown that each acrocentric chromosomes carries 30-50 ribosomal genes in a tandem array (Schmickel and Knoller 1977).

The organization can be seen best by direct EM visualization. Ribosomal DNA can be highly purified by CsCl centrifugation and R looping. The CsCl gradient takes advantage of their high buoyant density of 1.72 gm/ml compared to a buoyant density of 1.70 gm/ml of average DNA. If this DNA is hybridized to rRNA after CsCl enrichment, the DNA will gain even more buoyant density to permit purification. Figure 1 demonstrates a long rDNA strand that has hybridized to 18 and 28S rRNA. In 25 samples with two separate loops the distance between 18S or 28S loops was 43.85 kb (Sylvester et al. 1986).

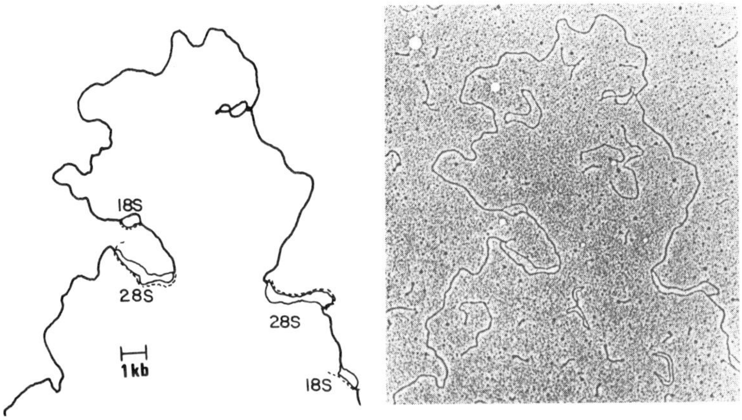

Figure 1. Electron micrograph and a tracing of r looped human DNA using rRNA.

The electron microscopy measurements are in good agreement with the reconstruction of the ribosomal gene by molecular cloning techniques. Wellauer and Dawid (1979) demonstrated that the genes could be characterized by four EcoRI fragments A, B, C, D. We have adopted their nomenclature with modifications. The A, B, C, D EcoRI fragments have been cloned with measurements of 5.6, 7.3, 12, 19 kb respectively to give a total length of 43.9 kb -- very close to that measured by EM (Wilson et al. 1978; Erickson et al. 1981; Sylvester et al. 1986). The 43.9 kb repeat is equal to 43,900 x 3.4A° or about 13 microns in length. The tandem complement of 30-50 genes vary from 400 to 700 microns in length. The EM measurements and the calculated length are much greater than the cytogenetically measured length and indicates that the rDNA genes must be rather tightly wound in nucleoli. The winding principle has not been determined.

The first ribosomal gene segment were isolated by the radioactive labeling of rRNA to identify lambda clones containing rDNA. The EcoRI A, B and D fragments contain the 18 and 28 S genes. Figure 2 gives the arrangement of these genes. The promoter for transcription located 700 bp upstream from the beginning of the mature 18S RNA, has been identified and sequenced by Financsek (1982) using the B clone. The transcription termination is not exactly known but preliminary data place it about 500 bp 3' to the end of the mature 28S RNA. Thus the rRNA transcript is 13 kb as shown in Figure 2 and includes an external transcribed spaced and two internal transcribed spacers that are removed in processing the primary transcript into the mature 18S, 5.8S and 28S RNAs.

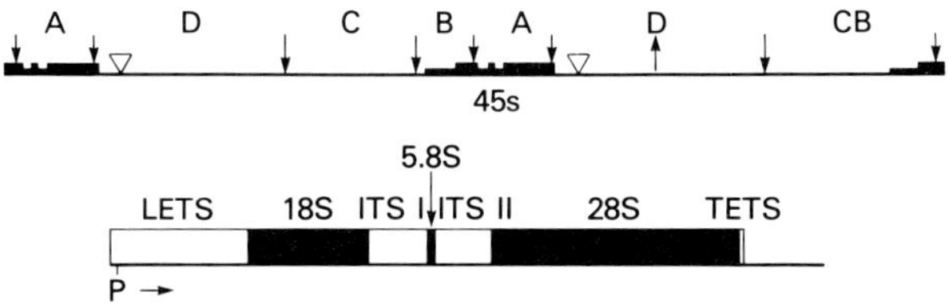

Figure 2. The complete ribosomal repeat (a) and the 45S primary transcript shown in detail (b). The filled blocks in (a) represent the primary transcript which includes the promoter P, the external transcribed spacer (LETS), the internal transcribed spacer (ITS I & II), and the terminal transcribed spacer (TETS).

The repeating pattern for the 30-50 ribosomal genes is a 13 kb transcript separated by a 31 kb spaced. The orientation of these two components in three dimensions is not known but could determine the accessibility of the genes for transcription.

Variability of rRNA genes. Are all ribosomal genes similar or are there variations? Restriction mapping of the transcribed portion shows only one variant, a HindII site in the 28S gene that is present in all primates. This is caused by a methylation difference in the DNA but no difference in DNA sequences. Just 5' to the promoter a variation of the EcoRI site is also found in great apes. By restriction mapping, the variation of the transcribed ribosomal genes is minimal.

In contrast to the lack of variation in the transcript, we note a consistent variation in the region just 3' to the transcript. This variation is revealed by BamHI digest and is varied in all individuals (Erickson et al. 1981). Furthermore, individual chromosomes do not show any specificity. Each chromosome contains several variants in size (Naylor 1983).

The BamHI variation is not an indication of mitotic instability since cloned cells do not show any change after extensive subculturing as documented in Hela cell lines. A given pattern is found in all tissues of an individual and for both of identical twins. In contrast, meiotic recombination is documented since new lengths and patterns are often seen in children that are not present in their parents.

The molecular basis of this variation is demonstrated best by cloning in lambda. Restriction analysis of lambda clones shows a series of tandemly repeated segments in some of the length variants. Those cloned variants that contain tandem repeats can and do generate size variants whereas those clones that do not contain a tandem repeat remain homogeneous during replication. The most likely explanation for this variation is unequal homologous exchange as seen in figure 3. This variability is an indication of an active recombination within each gene. Unequal homologous exchange between genes causes expansion and contraction of gene numbers within the tandem repeat in _Xenopus Laevis_ (Tartof 1974) and is probably the mechanism responsible for the generation of length diversity in the ribosomal gene stalks of acrocentric chromosomes in humans.

a)

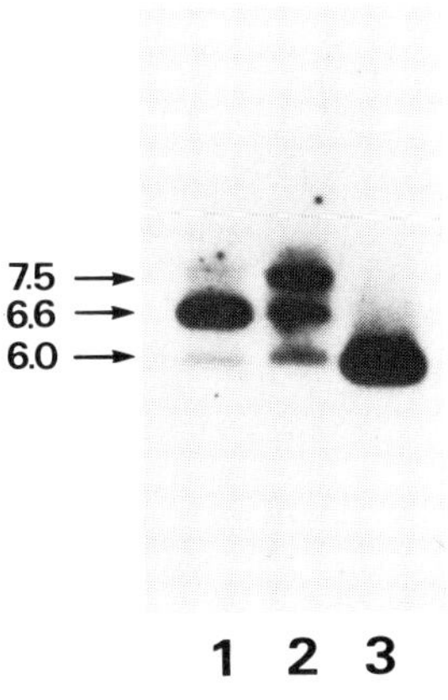

b)

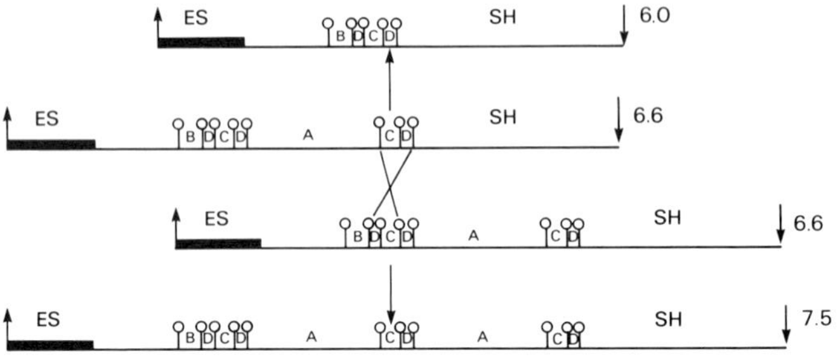

Figure 4. BamHI digest and Southern analysis of the variable spacer of rDNA 1. clone 6.6 fragment, 2. cloned genomic human, 3. clone 6.0 fragment (a) and a diagram of the unequal recombination that generates 6.0 and 7.6 fragments from 6.6 fragments. The lettered segments A, B, C, D are generated by restriction with Sal I. Recombination between the first and second set of C and D fragments generates unequal reciprocal products.

Other than the length heterogeneity noted at the 5' end of the primary transcript, we have identified a length heterogeneity at the 3' end of the gene in the C fragment. Considerable heterogeneity of the spacer has been found in the 9 kb HindIII fragment of the D fragment. However the extent and nature of that variation has not been carefully documented yet.

Sequence Information

The highly variable length of the 28 S ribosomal RNA among different species has been of interest. The 5025 bp human 28S RNA is almost 85% larger than that of E. coli (Gonzales et al. 1985). The 28S gene sequences have revealed an interesting relationship among all ribosomal sequences. All of the known ribosomes contain 12 invariant length tracts. Interspersed between these invariant tracts are variable sequences that are characteristic of kingdoms, phyla, classes and species. The invariant length regions share about 31% sequence homology between distantly related species such as E. coli and humans. However these regions show remarkable sequence homology between closely related species such as mouse and man (99%) and toad and man (96%). The factor responsible for the length constancy between man and E. coli is an extraordinary conservation of secondary structures. All of the secondary structures found in E. coli can be found in man. The authenticity of these structures is verified by compensatory base changes found in the proposed helices (Gorski et al. 1986). A compensatory base change is inferred when the mutation of a G→C is matched by a C→G change in the proposed pairing partner.

The variable tracts of ribosomal sequences are quite distinctive. The variable regions are 83% G+C as compared to the 59% G+C content of the constant tracts. The proposed structures for these are very long and stable double stranded helices. Often repetitive sequences are found within the variable sequences, $(GGC)_{11}$, $(CT)_8$. The variable regions may be specific for eucaryotes, vertebrates or mammals. There are eleven tracts defined as variable since they are not found in all organisms. We have analyzed the V5 sequence of human which shows no homology between mouse and rat or human. Between chimpanzee and human, there is a 6.1% difference and between gorilla and human there is a 7.1% difference. These regions evolve at a very high rate. When comparing the 1211 base pair constant region that flanks the variant region, the difference between man and chimpanzee is 0.08% instead of the 6.1% difference in the variant region. Thus, one part of the gene changes about seventy-five times more rapidly than a neighboring sequence.

The reason for differences in evolutionary rate is not readily apparent. We suspect that the variable regions are on the outside of the ribosome and are interactive with membranes or viruses. The variable regions may be subjected to selective forces such as attachment sites for invading viruses.

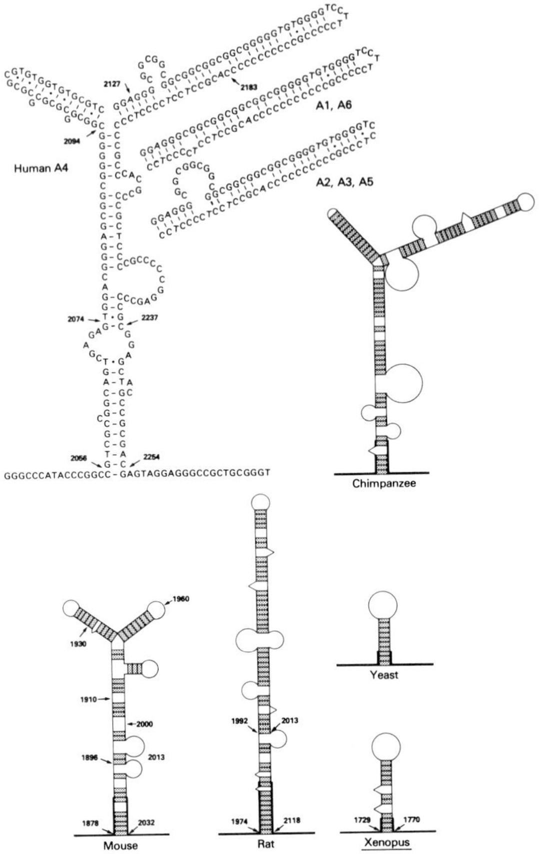

Figure 5. The primary and secondary structure of the V5 region of the human and chimpanzee V5 sequence. Six different human V5 regions were sequenced and the three different sequences are given. The relative length of the other known V5 regions are shown for rat, mouse, Xenopus and yeast. Each sequence gives a different secondary structure.

The variable sequences found in the 28S RNA are heterogeneous in humans. Probably no two ribosomes have the same sequence. We have begun to investigate diversity of ribosomal sequences by cloning and analyzing the human V5 region. Of 6 V5 regions examined, three different sequences were found. The different DNA sequences suggest a difference in rRNA. When a single DNA fragment is hybridized to cellular RNA and then digested with S1 nuclease to remove any mispaired DNA, the resulting heterogeneity reveals that cellular rRNA is a diverse population. We do not know if this sequence heterogeneity marks a functional diversity. However if no functional heterogeneity exists, the difference can be used to mark specific genes and measure the regulation of subsets of rRNA.

The 18S genes. The 18 S gene sequence is a much more conserved molecule than the 28S. The human 18S gene contains 1870 bp and is only 0.45% different than rat, 0.37 different than mouse and 2.6% different than Xenopus. Only 432 of the 1870 base pairs in 18S show variation among all species. When the 1438 bp which are conserved are considered the differences between man and mouse are only 0.1%. This is a remarkable level of conservation of a sequence over the 80 million years since the radiation of mammals. Certainly this level of conservation speaks to the important function of the 18S rRNA. However we know from comparative molecular studies that many base substitutions are permitted without a loss of function. Furthermore selection can not take place at the level of a single gene. The loss of one out of 400 ribosomal genes has no cellular effect. The extraordinary conservation of sequences is most likely the result of two processes that operate in concert. If recombination is frequent in a tandem set of genes, then there is the chance of correction of any "different" sequence during the formation of the hybrid double helices of the Holliday structure. Thus the organization of tandem genes will tend to preserve homogeneity in a tandem array. Occasionally the correction can serve to propagate a "different" sequence along a tandem series. If this happens then the cell may be at a disadvantage by the burden of a significant number of less functional genes at an otherwise active locus. Tandemness tends to retain homogeneity but when a homogeneous gene series has impaired function, selection can become a force for correction. Either tandemness or selection alone can not operate to conserve ribosomal sequences, but together these features result in the most highly conservation mechanism operating in evolution.

Significance of Sequence Studies

These sequence studies give us an interesting picture of the human ribosome, a patchwork of old and new. They also demonstrate that the ribosome is an ancient structure with a highly preserved core of subunits.

The marked difference between species and the lack of major differences within a species indicates that ribosomes on different chromosomes constitute a close interactive federation. Changes found between human and chimpanzee are not found between human ribosomal genes on different chromosomes. This is a different situation than that of related globin genes on different chromosomes whose counterparts are more closely related between species than between chromosomes. The α globin of chimpanzee and human are much more alike than the α and β globin genes.

The high rate of recombination between chromosomes which cause Robertsonian translocation is probably only a small fraction of all recombinations of the acrocentric chromosomes. If the ribosomal genes are oriented from telomere to centromere, as has been found in a 21 chromosome (Worton personal communication), it is probable that most exchanges are undetectable cytogenetically. Nevertheless the relatively large variation in ribosomal stalk size seen in metaphase chromosomes is evidence for recombination. Perhaps the best evidence for recombination between acrocentric chromosomes is the concerted evolution seen with ribosomal genes.

In extending these studies of ribosomal organization there are many principles yet to understand. The ones I have mentioned are: 1) the principle of winding and tertiary structure which must be related to gene activation, 2) the principles of interchromosomal recombination which can be better understood with careful family studies and specific chromosome markers, 3) the principles of gene variation which may reflect functional differences or serve as markers of subset regulation. The question of nondisjunction and its cause is hiding in the background somewhere and may be answered as we understand any of these principles.

References

1. Erickson, J.M., C.L. Rushford, D.J. Dorney, G.N. Wilson, R.D. Schmickel (1981) Structure and variation of human ribosomal DNA. Gene 16:1-9.

2. Financsek, I., K. Mizumoto, Y. Mishima, M. Muramatsu (1982) Human ribosomal RNA genes: Sequence for the transcription initiation region and comparison of three mammalian genes. Proc. Natl. Acad. Sci. USA 79:3092-3096.

3. Gonzales, I.L., J.L. Gorski, T.J. Campen, D.J. Dorney, J.M. Erickson, J.E. Sylvester, R.D. Schmickel (1985) Variation among human 28S ribosomal RNA genes. Proc. Natl. Acad. Sci. USA 82:7666-7670.

4. Gorski, J., I.L. Gonzales, R.D. Schmickel (1986) The secondary structure of human 28S rRNA: the structure and evolution of a mosaic rRNA gene. J. Mol. Evol., in press.

5. Henderson, A.S., D. Warburton, K.L. Atwood (1972) Location of rDNA in the human chromosome complement. Proc. Natl. Acad. Sci. USA 69:3394-3398.

6. Magenis, R.E. and J. Chamberlin (1981) Parental origin of nondisjunction in Trisomy 21. F. de la Cruz and P.S. Gerald eds., 77-93, University Park Press, Baltimore.

7. Naylor, S.L., A.Y. Sakaguchi, R.D. Schmickel, M. Woodworth-Gutai, T. Shows (1983) Organization of rDNA spacer fragment variant among human acrocentric chromosomes in somatic cell hybrids. J. Mol. Appl. Genet. 2:1377-146.

8. Schmickel, R.D. and M. Knoller (1977) Characterization and localization of the human genes for ribosomal RNA. Pediatr. Res. 11:929-935.

9. Stahl, A. (1982) The nucleolus and the nucleolar chromosomes. In The Nucleolus E.G. Jordan and C.A. Cullis, eds. Cambridge University Press, Cambridge.

10. Sylvester, J.E., D.A. Whiteman, R. Podolsky, J.M. Pozsgay, J. Respess, R.D. Schmickel (1986) The human ribosomal gene. Hum. Genet., in press.

11. Tartof, K.D. (1974) Unequal mitotic sister chromatid exchange as the mechanism for ribosomal gene magnification. Proc. Natl. Acad. Sci. USA 71:1272-1276.

12. Wellauer, P., I.B. Dawid (1979) Isolation and sequence organization of human ribosomal DNA. J. Mol. Biol. 98:503-517.

13. Wilson, G.N., B.A. Hollar, J.R. Waterson, R.D. Schmickel (1978) Molecular analysis of cloned human 18S ribosomal DNA segments. Proc. Natl. Acad. Sci. USA 75:5367-5371.

Nucleolus organizer regions and nucleoli: cytological findings

H. G. Schwarzacher and F. Wachtler

Histologisch-Embryologisches Institut der Universität Wien, Schwarzspanierstrasse 17, A–1090 Wien, Austria

For a better understanding of the nucleolar structures it is advisable to make a distinction between the nucleolus organizer region and the nucleolus:

The nucleolus organizer region (NOR) is defined as a region of a chromosome (containing the major rRNA-genes), hence it is made up of the chromatin fibril.

The nucleolus is a structure which is formed whenever the NOR is active, or has been active shortly before, and presumably also when it is kept ready for action.

There are usually several NORs per genome, but their number may differ among species (in primates e.g. between 1 and 9 per haploid genome). In each NOR many rDNA-transcription units are repeatedly aligned. The number of rDNA-transcription units per genome is within one class or one genus relatively constant (in mammals e.g. about 200), thus the number of rDNA-transcription units per NOR varies between species (see Hadjiolov for review). Moreover individual differences in size and activity of NORs can be found (see Schwarzacher and Wachtler,1983,for review).

A nucleolus can be formed by either one or by more than one NORs, it contains of course the NOR(s) but besides also the transcription product and the machinery to make ribosomes. With that it is an unique structure within the cell nucleus combining transcription, processing of the transcription product, and the assembling of cell organelles. We may therefore expect the following structures within the nucleolus:

NOR(s), chromatin containing repeated DNA units of pre-rRNA genes and non-transcribed spacer; Histones, and non-histone chromosomal proteins; Primary transcripts (45S-RNA); Intermediate rRNA species resulting from processing 45S-RNA; Final rRNAs (18S-, 5.8S-, 28S-rRNA, and 5S-rRNA transported to the nucleolus from other parts of the genome; Proteins associated with pre-rRNAs and rRNAs; Preribosomes (80S, 55S, 40S); L- and S-ribosomal subparticles; Enzymes for transcription, processing of RNA, and preribosome formation; Matrix (still in discussion).

The electron microscopic studies (for reviews see Smetana and Busch 1974; Stahl,1982; Schwarzacher and Wachtler,1983; Goessens,1984) made it clear that in the nucleoli mainly three components are to be found (fig.1):

1. Small areas of a fine fibrillar material giving low electron contrast and situated in most cases well inside the nucleolus, and therefore called "fibrillar centers".

2. Layers of fibrillar material of high electron contrast, called "dense fibrillar component", in some nucleoli surrounding the fibrillar centers (fig.1), in others forming a network (known also as "nucleolonema", fig.2).

3. Small granules, in most cases situated more at the periphery of the nucleolus, called "granular component".

Sometimes occurring conspicuous small areas, mainly filled with low-contrast chromatin and with perichromatin granules, called "nucleolar vacuoles" should perhaps not be interpreted as a component of the nucleolus but rather as the sign of a reticulate structure of the nucleolus intermingled with non-nucleolar components of the nucleus. Also the condensed chromatin known as "nucleolus associated heterochromatin" should not be regarded as a real nucleolar component. Its existence differs very much between species, and any direct connection with nucleolar function is not known so far.

Nucleoli may change very much in their form and inner structure but display nevertheless in general the three components described above. This is true for many plant and animal species.

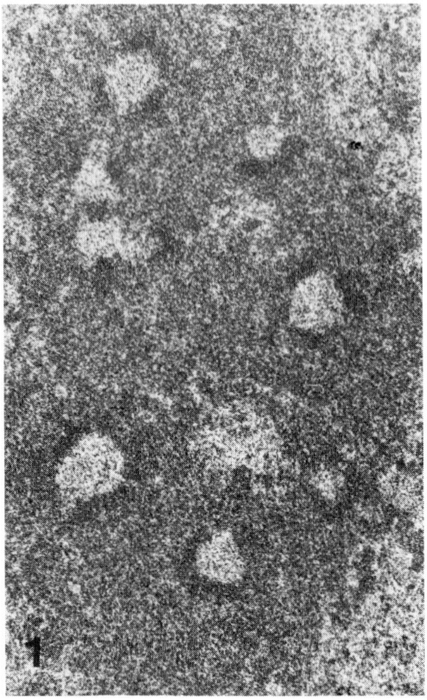

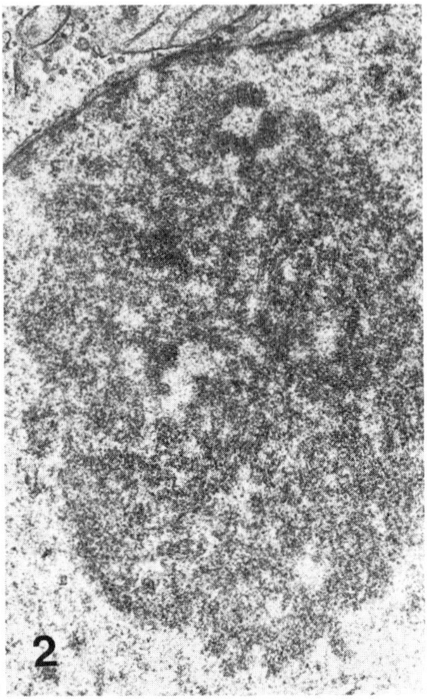

Fig.1 : Large compact nucleolus of a human melanoma cell with several fibrillar centers. (Magnification 22.000x).

Fig.2 : Nucleolus of the nucleolonema type of a human fibroblast in vitro. The dense fibrillar component forms a network, fibrillar-center-regions are sparse.(Magnification 22.000x)

In this short review the mammalian and particularely the human nucleoli are taken as examples. Different types and forms of nucleoli may for instance develope with different states of ribosome biosynthesis. This can be illustrated by human lymphocytes (see e.g. Wachtler et al., 1980):

Small, non-stimulated lymphocytes of the peripheral blood contain in most cases only one small nucleolus with one fibrillar center ("ring-shaped" nucleolus, in light microscopic silver-preparations seen as small dot, see fig.3). With stimulation (e.g. by phytohemagglutinine) the dense fibrillar component increases very much during the first 6 - 12 hours, forming a network (nucleolonema). Similar structured nucleoli are found in regenerating cells or in very rapidly dividing cells (e.g. fibroblasts in vitro, fig.2). In fully activated lymphocytes (blast-like cells) large compact nucleoli with several fibrillar centers (as in fig.1) are found.

A matter of much debate has been the exact localization of the NOR within the nucleolus.

DNA has been demonstrated mainly in the fibrillar centers and in the dense fibrillar component. Labelling experiments using ^{3}H-uridin showed that in most instances RNA synthesis takes place in or very near the dense fibrillar component. The granular component is progressivly labelled after longer time intervals indicating the newly synthetized RNA to be shipped into the granular component towards the periphery of the nucleolus (see reviews by Stahl,1982; Fakan and Puvion, 1982; Goessens,1984). These observation lead to the view that transcribed rDNA is situated in the dense fibrillar component, and that the granular component contains the preribosomes.

Considerable interest found the fact that fibrillar centers and, to a somewhat lesser degree, the dense fibrillar component contain acidic proteins which can be stained with silver salts (see reviews by Schwarzacher and Wachtler, 1983; Goessens, 1984). The silver staining is responsible for the demonstration of the NORs in metaphase spreads. But only those NORs that were actively transcribed in the preceeding interphase show a positive silver reaction (Miller et al.,1976). Therefore the most likely interpretation of the silver staining of NORs in metaphase is that a remnant of the nucleolus remains attached to them. This is clearly shown in electron micrographs of whole mounted silver stained chromosomes (Schwarzacher et al.,1978). Furthermore, fibrillar-center-like material can be detected attached to mitotic chromosomes in electron micrographs of ultrathin sections(see Schwarzacher and Wachtler,1983).

It has been repeatedly discussed if the fibrillar centers are identical with the NORs ? As the above mentioned silver staining demonstrates, there must be much more included in the fibrillar centers than just the chromatin fibril. Indeed several non-chromosomal proteins have been found within the fibrillar centers (see Hadjiolov, 1985, for review). U.Scheer reported in this conference that RNA-polymerase I (demonstrated by electron microscopic immune histochemistry) is situated in the fibrillar centers, indicating rDNA-transcription at this site. This is somewhat in contradiction to the ^{3}H-uridine labelling experiments (see above) indicating the site of rDNA-transcription to be restricted rather to the dense fibrillar component.

Another important finding is the lack of nucleosomes in chromatin spreading preparations of fibrillar centers (see Derenzini et al., this conference). It must be remembered, however, that not only transcribed rDNA units but also the non-transcribed spacers are devoid of nucleosomes.

At present we have therefore indications that rDNA transcription takes place in the dense fibrillar component and/or the fibrillar centers. A certain amount of dense fibrillar component is found in all types of nucleoli. Fibrillar centers, however, may vary very much with the activity of the cell, and they are not formed in all types of nucleoli. If fibrillar centers are formed within a nucleolus, they may well contain also the rRNA genes.

Inactive NORs should not participate in the formation of a nucleolus. Therefore they cannot be generally related to fibrillar centers or the dense fibrillar component. This is clearly shown in inactive cells with no or only a small nucleolus:

Silver stained resting lymphocytes display in most cases only one small silver-positive nucleolus containing a single fibrillar center (fig.3). Human diploid cells contain, however, 10 NORs. Preparations demonstrating a in situ hybridization of fluorescence labelled rRNA to rDNA (fig.3, right) show in each cell nucleus several, sometimes up to 10 rDNA spots (Wachtler et al.,1986). This indicates an occasional fusion of some of the NORs, but never a fusion to only one nucleolus. If each NOR were connected with a fibrillar center, so many silver dots as rDNA spots should be seen. This was never found in our investigations.

These observations underline the view that fibrillar centers and the surrounding dense fibrillar component in form of a shell are special structural organizations which develope in certain active cells with a high rate of ribosome biosynthesis.

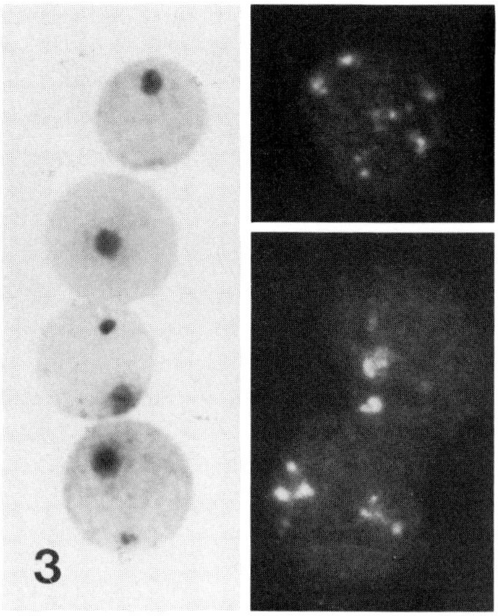

Fig.3 : Nuclei of non-stimulated lymphocytes from peripheral blood.

Left: Ag-staining, usual one nucleolus per cell nucleus.

Right: In-situ-hybridization (rRNA - rDNA). Several spots (fluorescence signals) indicating position of NORs.

Magnification 2000x

Fibrillar centers may not always and not only be directly connected with transcription, but may also function as a stock or a store for the proteins needed for transcription, RNA processing and ribosome formation. This is suggested by experiments employing inhibition of RNA synthesis and of protein synthesis.

In a series of experiments in our laboratory (in cooperation with W.Popp) we found that in several different human cell lines, including lymphocytes, fibroblasts, Hela cells, melanoma cells) the inhibition of protein synthesis by puromycin leads to a marked decrease of the volume of fibrillar centers whenever the cells are actively growing and synthetizing ribosomes. In non-active cells, as e.g. in resting lymphocytes, the fibrillar centers remain practically of the same size. This indicates a turnover of the proteins of the fibrillar centers which is dependent on transcription and preribosome assembling. The inhibition of RNA-synthesis by actinomycine does not affect the size of the fibrillar centers, but of course the amount of the granular component, as was already shown previously by several authors (e.g. Jordan and Mc Govern,1981; Lepoint and Goessens, 1985).

These findings suggest that in small resting lymphocytes the single nucleolus keeps a distinct fibrillar center as a store of proteins in order to be able to respond quickly to the demand of ribosome formation in the moment of stimulation by an antigene. The fibrillar center and the surrounding dense fibrillar component may additionally serve as a functional unit containing the few transcribed rDNA-units.

The nucleoli with nucleolonema, containing only few fibrillar-center material, much dense fibrillar component, and relatively few granular component are found in lymphocytes a few hours after stimulation, in fast dividing, and in regenerating cells (fig.2). In freshly stimulated cells it can be supposed that the so far not transcribed NORs are now activated, forming a network consisting of the dense fibrillar component. The fibrillar center-material is reduced because it is used up and not yet restored. Ribosomes have to be made first very fast to synthetize the proteins which are in turn necessary for a further ribosomal assembling; the granular component is rather sparse in such nucleoli because preribosomes are shipped to the cytoplasm at a high rate. The nucleolonema-type nucleoli show that the transcribed rDNA needs not necessarily be situated in fibrillar centers. A much better candidate for the localization of rDNA in such nucleoli is the network (nucleolonema) of the dense fibrillar component.

The large nucleoli with several or many fibrillar centers (fig.1) are found in steadily active cells such as secretory cells, nerve cells, fully activated lymphocytes. Multiple functional units of rDNA plus the necessary proteins in store are here formed in the shape of fibrillar centers and surrounding shells of dense fibrillar material. The preribosomes are accumulated around these centers composing the vast granular component. The relations between the different nucleolar components remain now stable in such nucleoli. Since the fibrillar centers may by far outnumber the NORs (Jordan and McGovern,1981), each such functional center may contain only parts of one NOR. On the other hand, each center may contain several transcription units (Mirre and Knibiehler, 1984).

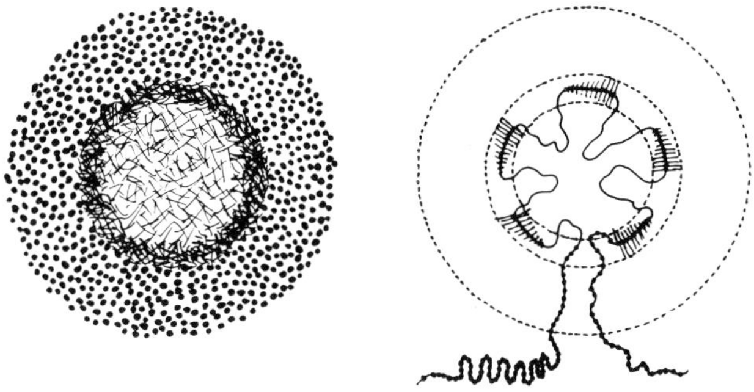

4

Fig.4 : Model of a "nucleolar unit" consisting of a fibrillar center surrounded by a shell of dense fibrillar component and towards the outside by the granular component.

Left : Schematic drawing of the electron microscopic view.

Right : Course of the NOR through the nucleolar unit. Transcription units drawn as "christmas trees" within the dense fibrillar component, non-transcribed spacer DNA as loops reaching into the fibrillar center. Only 5 transcription units drawn.

In fig.4 a possible model of such a "nucleolar unit", as it is observed in the small "ringshaped" nucleoli and in the large compact nucleoli (containing several centers) is given. The exact position of the transcribed rDNA and the non-transcribed spacer is not yet clear, as has been discussed above. In this model we follow Stahl (1982) and Hadjiolov (1985), placing the transcribed rDNA in the dense fibrillar component and the non-transcribed spacer DNA in the fibrillar center.

REFERENCES

Fakar, S. and S. Puvion 1980. Ultrstructural visualization of nucleolar and extranucleolar RNA-synthesis and distribution. Int. Rev. Cytol. 65, 255-299.

Goessens, G. 1984. Nucleolar structure. Int.Rev.Cytol. 87, 107-158.

Hadjiolov, A.A. 1985. The nucleolus and ribosome biogenesis. Wien and New York: Springer.

Jordan, E.G. and McGovern, J.H. 1981. The quantitative relationship of the fibrillar centers and other nucleolar components to changes in growth conditions, serum deprivation and low doses of actinomycin D in cultured diploid human fibroblasts (strain MRC 5). J.Cell Sci. 52, 373-389.

Lepoint, A. and G.Goessens 1985. Effect of 2-mercapto-1(ß-4-pyridethyl) benzimidazole (MPB) on Ehrlich cell nucleoli: Stereological analysis. Biol. Cell 53, 111-116.

Miller, D. A., V.G. Dev, R. Tantravahi and O. J. Miller 1978. Suppression of human nucleolus organizer activity in mouse-human somatic hybrid cells. Exp. Cell Res. 101, 235-243.

Mirre, C. and B. Knibiehler 1984. Quantitative ultrastructural analysis of fibrillar centers in the mouse. Correlation of their number and volume with nucleolar organizer activity. Protoplasma 121, 120-128.

Schwarzacher, H. G., A. V. Mikelsaar and W.Schnedl 1978. The nature of Ag-staining of nucleolus organizer regions. Cytogenet. Cell Genet. 20, 24-39.

Schwarzacher, H. G. and F. Wachtler 1983. Nucleolus organizer regions and nucleoli. Hum. Genet. 63, 89-99.

Smetana, K. and H. Busch 1974. The nucleolus and nucleolar DNA. In The cell nucleus vol.1, H. Busch, ed., 75-147. New York and London: Academic press.

Stahl, A. 1982. The nucleolus and nucleolar chromosomes. In The nucleolus, Jordan, E. G. and C. A. Cullis eds., 1-24. Cambridge: Cambridge University Press.

Wachtler, F.,A. Ellinger and H. G. Schwarzacher 1980. Nucleolar changes in human PHA-stimulated lymphocytes. Cell Tiss. Res. 213, 351-360.

Wachtler, F., A. H. N. Hopman, J. Wiegant and H. G. Schwarzacher 1986. On the position of nucleolus organizer regions (NORs) in interphase nuclei. Studies with a new, non-autoradiographic in situ hybridization method. Exp. Cell Res. , in press.

Aknowledgement: The original investigations reported in this paper were supported by the Österreichische Fonds zur Förderung der wissenschaftlichen Forschung (Projekt P 5665).

Relations between nucleoli and nucleolus-organizing regions during the cell cycle

G. Goessens, M. Thiry and A. Lepoint

Institute of Histology, University of Liège, B–4020 Liège, Belgium

A clear relationship between nucleolus and some chromosomal regions was first established by Heitz (1931). This author demonstrated the involvement of these chromosomal regions in the telophase reconstitution of the nucleoli. McClintock (1934) called these regions "nucleolus organizing regions" (NORs), a term indicating that the nucleoli originate in these regions.

In metaphase chromosomes, the NORs appear in many species as particularly clear constrictions when the chromosomes are stained with Feulgen reaction: the secondary constrictions. This fact may be due to the low degree of condensation of DNA or to the appreciable protein content in the secondary constrictions (Busch and Smetana, 1970).

In modern terms, the NORs can be specified as chromosomal locations of the main rRNA genes (Ritossa and Spiegelman, 1965; Wallace and Bernsteil, 1964). The most direct method of locating the NORs is by in situ hybridization using ^{3}H-RNA hybridized directly to cytological preparations. Autoradiographic silver grains have been observed in correspondence with the secondary constrictions (Henderson et al., 1972; Hsu et al. 1975). An other method of identifying NORs is by using silver staining procedure (Howell et al., 1975; Goodpasture and Bloom, 1975). The silver stained regions can be equated with the location of 18 + 28S rRNA cistrons.

Electron microscopic studies of secondary constrictions have shown that the achromatic regions are not really constrictions (Hsu et al., 1967; Lafontaine, 1968; Brinkley and Stubblefield,

Chromosomes today, volume 9 Allen & Unwin 1987

1970; Goessens and Lepoint, 1974). The width of the NORs is approximatively the same as that of the chromosome arms. In fact, this constriction predominantly consists of fibrillar material which following classical stainings exhibits a low electron density contrasting with the adjoining chromosome segments. However, after silver staining procedure, the NORs are positively contrasted.

During interphase, the ribosomal genes are transcribed and the nucleoli differentiate around the NORs.

In the last past years, numerous papers and some reviews have been published concerning the nucleolus and its relations to the NORs. However, as mentionned by Schwarzacher and Wachtler (1983), several publications still reveal differences in the interpretation of those results and the identification of NORs during interphase among the nucleolar components remains controversial.

For these reasons, we shall summarize some cytological and cytochemical ultrastructural results and problems with particular reference to the example of Ehrlich tumour cells which are very convenient model to study the behaviour of the nucleolus during the cell cycle and its relations to NOR.

In the first part of this paper, we summarize the behaviour of the nucleoli during interphase and mitosis. In the second one, we analyze the ultrastructure of the NORs in metaphase chromosomes.

THE NUCLEOLUS DURING INTERPHASE AND MITOSIS

During interphase

At least five nucleolar components can be recognized: the fibrillar centre(s), the dense fibrillar component, the granular component, the nucleolar vacuoles or interstices and the nucleolus-associated chromatin (Fig. 1 - for reviews see Smetana and Busch, 1974; Fakan, 1978; Puvion and Moyne, 1981; Jordan and Cullis, 1982; Hernandez-Verdun, 1983; Goessens, 1984; Fakan and Hernandez-Verdun, 1986).

Fibrillar centres are areas of low electron density and fibrillar in texture. In active nucleoli, they are surrounded more or less completely by dense fibrillar component. It is generally admitted that fibrillar centres contain some DNA which is connected with condensed intranucleolar chromatin (Goessens, 1979; Ashraf and Godward, 1980). The chromatin in the fibrillar centres is in a non-nucleosomal configuration as revealed by the Feulgen-like osmium-ammine technique (Derenzini et al., 1983). In situ hybridization performed at the photonic level in human spermatocytes (Arroua et al., 1982) demonstrated that, in this material, the fibrillar centres contain rDNA. However, transcription of the ribosomal genes essentially occurs at the periphery of the fibrillar centres in the layer of dense fibrillar component surrounding them (Goessens, 1976; Mirre and Stahl, 1978 - review in: Fakan, 1978; Stahl, 1982; Goessens, 1984). The fibrillar centres mainly contain proteins specially silver stained proteins also localized in the dense fibrillar component (Fig. 2). Several studies suggested that nucleolar argyrophilic proteins correspond to phosphoprotein C23 (Ochs and Busch, 1984; Smetana et al. 1984), to a 36Kd nucleolar protein (Hubbell et al., 1979) or to the large RNA polymerase I subunit (Williams et al., 1982). In addition, Scheer and Rose (1984) showed by immunocytochemistry that fibrillar centres contain RNA polymerase I.

Dense fibrillar component is the site of pre-rRNA synthesis. This dense fibrillar component containing rapidly labelled RNA can be visualized not only around but also inside the fibrillar centres (Thiry et al., 1985). In addition to newly synthesized RNA, the dense fibrillar component contain DNA and silver stained proteins (Fig. 2).

Granular component. As the fibrils, the granules are composed of RNP. Ultrastructural, autoradiographic and biochemical data demonstrated that the nucleolar granules are formed in close proximity to the fibrillar component, later to be released as preribosomal particles and finally transported to the cytoplasm as ribosomal subunits (review in Hadjiolov, 1985).

Thus the two RNP nucleolar components namely the fibrillar component and the granular component relate to two stages in the formation of ribosomes.

Nucleolar vacuoles. The morphology, the size and the number of nucleolar vacuoles are variable and depend on the nucleolar type. They generally contain nucleoplasmic material. The possible function of nucleolar vacuoles could be the increase in the nucleolus-nucleoplasm exchange interface (Deltour and De Barsy, 1985).

Nucleolus-associated chromatin. Some condensed chromatin is found adjacent to the nucleolus (the peri-nucleolar chromatin) and penetrating among the other nucleolar components (the intranucleolar chromatin). The fibrillar centres have been shown to be attached to and continuous with the intranucleolar condensed chromatin (Goessens, 1979; Ashraf and Godward, 1980; Thiry and Goessens, 1986).

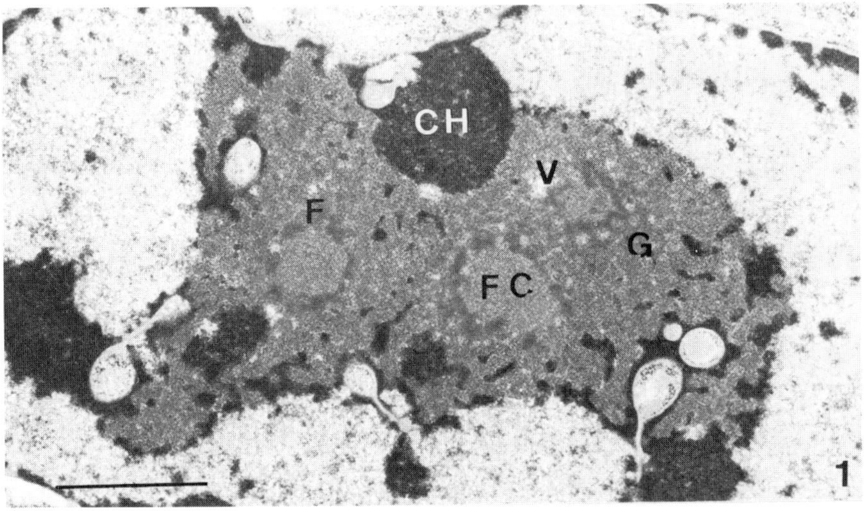

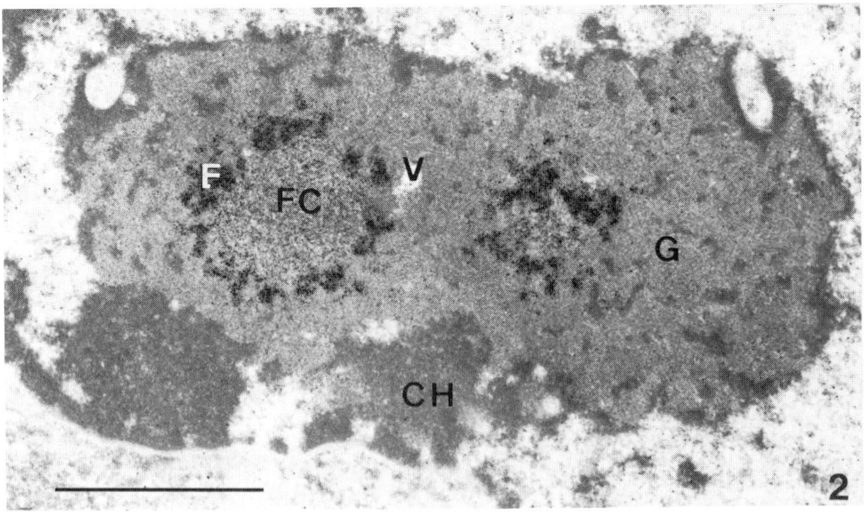

FIGURES 1 and 2. Ehrlich tumour cell nucleoli. Fig. 1: after acetylation method. Fig. 2: after silver and acetylation methods. FC: fibrillar centre, F: dense fibrillar component, G: granular component, V: nucleolar vacuole, CH: nucleolus-associated chromatin. Bars = 1 μm.

Nucleolar morphology varies widely depending on cell type and physiological state of the cell. In fact, it depends on the balance among the rRNA synthesis, the production and the accumulation of nucleolar proteins and the transport of ribosomal subunits out of the nucleolus (Goessens, 1976, 1978, 1984).

A combination of several techniques including velocity sedimentation, autoradiography, cytophotometry and stereology allowed quantitative data about the growth of the nucleolus and different nucleolar components in the course of interphase in Ehrlich tumour cells (Lepoint and Goessens, 1982). It was observed that the nuclear and nucleolar volumes double during interphase and that all the nucleolar components participate in the doubling of the nucleolar volume during interphase.

However, nucleoli of various cell types do differ in the percentage of the total nucleolar area occupied by each ultrastructural component. The significance of these differences is probably related to differences in rate of RNA synthesis and maturation.

During mitosis

The behaviour of the nucleoli during mitosis has been studied in a lot of cellular species and among others in Ehrlich tumour cells (review in Goessens, 1984).

During prophase, the changes in shape and the gradual reduction in size of the nucleolus is accompanied by the disappearance of the dense fibrillar component and the dispersion of the granular component.

During metaphase and anaphase, the fibrillar centres are detectable associated with some chromosomes (Fig. 3) and are also positive, as the interphasic fibrillar centres are, after silver staining (Fig. 4). These silver positive structures get a spherical or a more elongated and curved shape. In some section, the argyrophilic structures are associated in pairs (Fig. 4) and are localized in dips diametrically opposed on the edge of chromosomes (Thiry, 1986; Ploton et al., 1986).

During telophase, the fibrillar component reappears in close association with the fibrillar centres. Shortly afterwards, the granular component is also present and the nucleolus progressively reacquires its interphase ultrastructure (Fig.5).

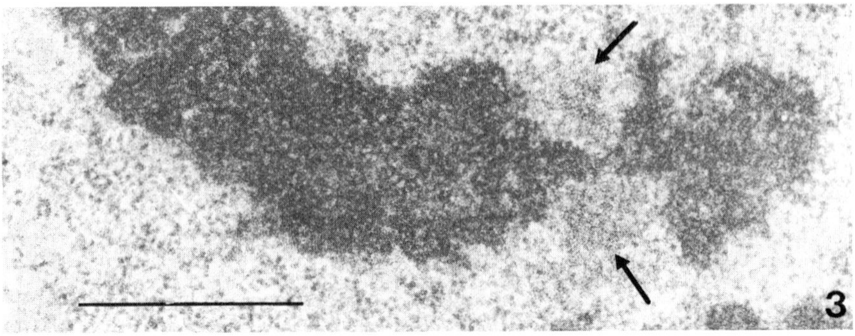

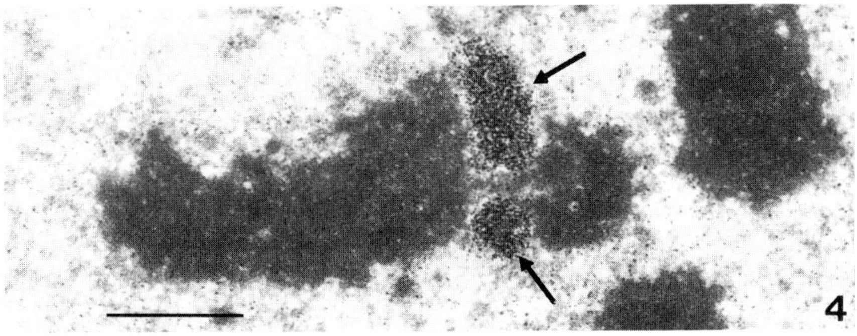

FIGURES 3 and 4. Sections from metaphase chromosomes after classical staining (Fig. 3) or after silver and acetylation methods (Fig. 4). Argyrophilic structures corresponding to fibrillar centres (arrows) are associated in pairs with chromosomes. Bars = 0,5 μm.

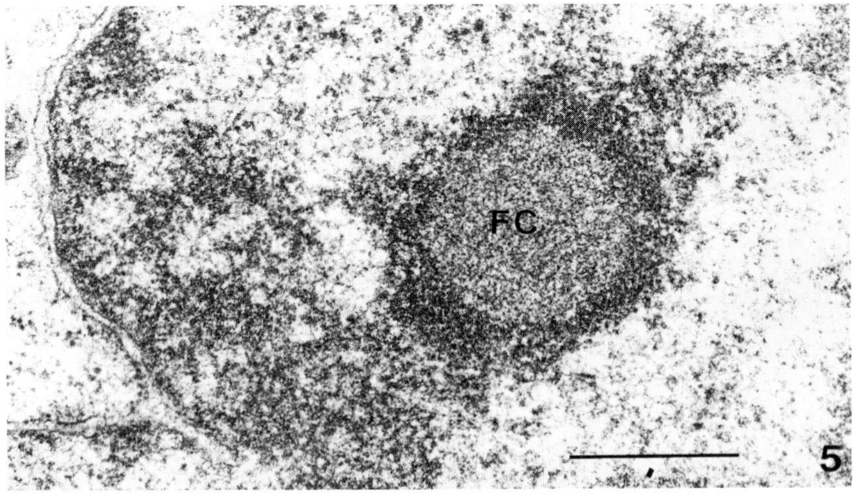

FIGURE 5. Nucleologenesis during telophase. Around the fibrillar centre (FC), the other nucleolar components progressively reappear during telophase. Bar = 0,5 μm.

On the basis of their morphology, their silver staining properties and their behaviour during mitosis, it is clear that the fibrillar centres correspond, at least in part, to NORs.

NOR ULTRASTRUCTURE IN METAPHASE CHROMOSOMES

As previously mentionned, the NORs appear, in ultrathin sections of metaphase chromosomes, as areas of low electron density (Fig.3). These areas contain silver stained proteins (Fig. 4) and can be visualized included or associated with chromosomes. The NORs could be considered as chromosomal regions with a tendency of incomplete condensation giving rise to the lightly staining zones resembling fibrillar centres of nucleoli.

However, according to Schwarzacher and Wachtler (1983), the electron microscopic analysis of a total preparation of silver stained metaphase chromosomes demonstrated that silver positive material is located around and beside the chromosomes but never directly in the chromosomal material. Thus, these observations point to the persistence of at least parts of nucleolar fibrillar centres through mitosis; this material is attached to the NOR but is not the NOR itself (Schwarzacher and Wachtler, 1983).

In order to clarify this problem, serial sections were performed in Ehrlich metaphase chromosomes after combined Ag-NOR and acetylation method (Thiry, 1986). From these data, a model of the secondary constriction structure can be proposed (Fig. 6). In metaphase, the nucleolar chromosomes carry a pair of argyrophilic doublets, one doublet on each chromatid. These two spherical structures are localized in dips diametrically opposed on the edge of chromatids. These spheres are partly integrated within the chromosomes as already pointed out by other authors (Paweletz and Risueno, 1982; Ploton et al., 1985, 1986). The various ultrastructural pictures obtained after sections of chance can easily be explained by different incidences of sections in this model as illustrated in Fig. 6.

It is of great importance to mention that, in addition to nucleolar proteins, these areas of low electron density contain DNA. This DNA present at the sites of silver granule localization has a completely extended configuration as demonstrated by Hernandez-

Verdun and Derenzini (1983).

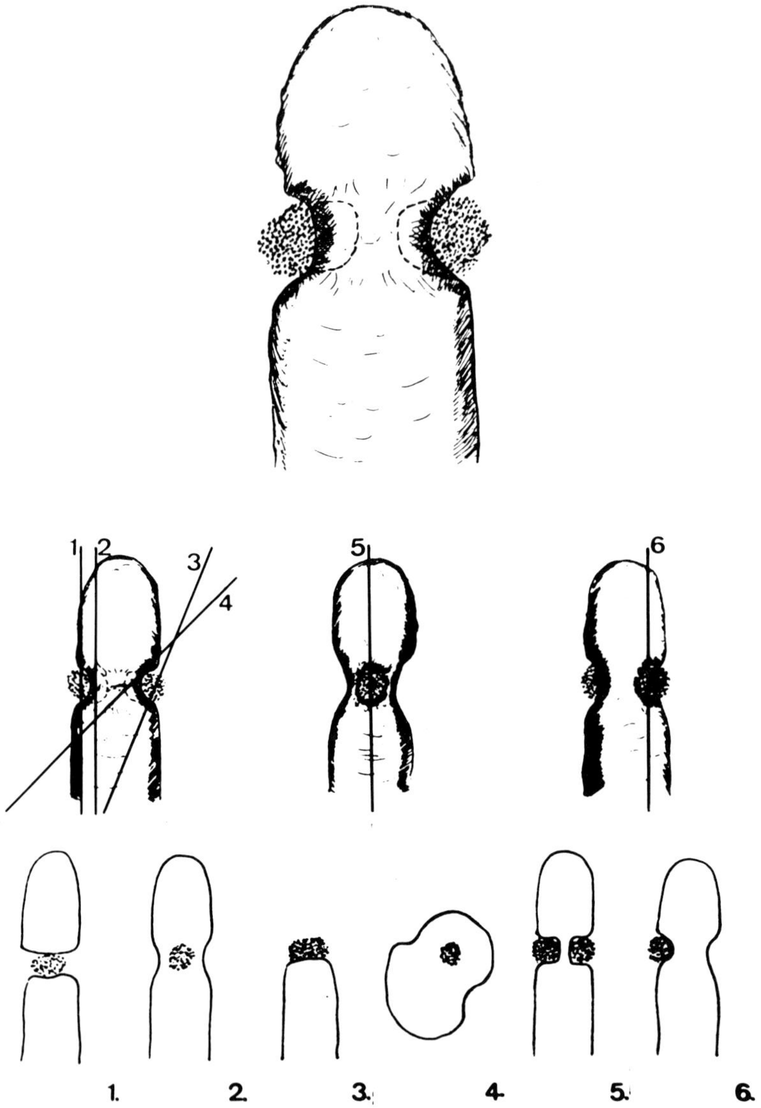

FIGURE 6. Above: Schematic representation of a secondary constriction in Ehrlich metaphase cell. A portion of one chromatid is illustrated. The latter carries a pair of argyrophilic structures. These two argyrophilic structures are situated in dips (discontinuous lines) diametrically opposed on chromatid.

Below: Interpretation of ultrastructural pictures obtained after various sections (continuous lines) in this model. This figure is reproduced from Thiry (1986) with permission by Bulletin de la Société Royale des Sciences de Liège.

CONCLUSION

There is now much evidence in support of the view that fibrillar centre are, at least in part, counterparts of NORs.

Acknowledgments: We thank Francine Skivée for skillfull technical assistance. This work was supported by grants from the "Fonds de la Recherche Scientifique Médicale" (n° 3.4512.86) and from the Belgian Government (Action de Recherche Concertée n° 85/90-80).

REFERENCES

Arroua, M.L., M. Hartung, M. Devictor, J.L. Bergé-Lefranc and A. Stahl. 1982. Localisation of ribosomal genes by in situ hybridization in the fibrillar centre of the nucleolus in the human spermatocyte. Biol. Cell 44, 337-340.

Ashraf, M. and M.B.E. Godward. 1980. The nucleolus in telophase, interphase and prophase. J. Cell Sci. 41, 321-329.

Brinkley, B.R. and E. Stubblefield. 1970. Ultrastructure and interaction of the kinetochore and centriole in mitosis and meiosis. In Advances in cell biology, vol. 1, D.M. Prescott, L. Goldstein and E. McConkey, ed., 119-185. New York: Academic Press.

Busch, H.and K. Smetana. 1970. The Nucleolus. New York: Academic Press.

Deltour, R. and T. De Barsy. 1985. Nucleolar activation in embryo radicle cell during early germination. J. Cell Sci. 76, 67-83.

Derenzini, M., A. Pession, Ch.M. Betts-Eusebi and F. Novello. 1983. Relationship between the extended, non-nucleosomal intranucleolar chromatin in situ and ribosomal RNA synthesis. Exp. Cell Res. 145, 127-143.

Fakan, S. 1978. High resolution autoradiography studies on chromatin functions. In The cell nucleus, vol. 5, H. Busch, ed., 3-53. New York: Academic Press.

Fakan, S. and D. Hernandez-Verdun. 1986. The nucleolus and the nucleolar organizer regions. Biol. Cell 56, 189-206.

Goessens, G. 1976. The nucleolar fibrillar centres in various cell types in vivo and in vitro. Cell Tiss. Res. 173, 315-324.

Goessens, G. 1976. High resolution autoradiographic studies of Ehrlich tumour cell nucleoli. Nucleolar labelling after ^{3}H-actinomycin D binding to DNA or after ^{3}H-TdR or ^{3}H-uridine incorporation in nucleic acids. Exp. Cell Res. 100, 88-94.

Goessens, G. 1978. Nucleolar ultrastructure during reversible inhibition of RNA synthesis in chick fibroblasts cultivated in vitro. J. Ultrastr. Res. 65, 83-89.

Goessens, G. 1979. Relations between fibrillar centres and nucleolus-associated chromatin in Ehrlich tumour cells. Cell Biol. Int. Rep. 3, 337-343.

Goessens, G. 1984. Nucleolar structure. Int. Rev. Cyt. 87, 107-158.

Goessens, G. and A. Lepoint. 1974. The fine structure of the nucleolus during interphase and mitosis in Ehrlich tumour cells cultivated in vitro. Exp. Cell Res. 87, 63-72.

Goodpasture, C. and S.E. Bloom. 1975. Visualization of nucleolar organizer regions in mammalian chromosomes using silver staining. Chromosoma 53, 37-50.
Hadjiolov, A.A. 1985. The nucleolus and ribosome biogenesis. Wien, New York: Springer-Verlag.
Heitz, E. 1931. Die Ursache der gesetzmässignen Zahl, Lage, Form und Grösse pflanzlicher Nukleolen. Planta 12, 775-844.
Henderson, A.A., D. Warburton and K.C. Atwood. 1972. Location of ribosomal DNA in the human chromosome complement. Proc. Natl. Acad. Sci. USA. 69, 3394-3398.
Hernandez-Verdun, D. 1983. The nucleolar organizer regions. Biol. Cell 49, 191-202.
Hernandez-Verdun, D. and M. Derenzini. 1983. Non-nucleosomal configuration of chromatin in nucleolar organizer regions of metaphase chromosomes in situ. Europ. J. Cell Biol. 31, 360-365.
Howell, W.M., T.E. Denton and J.R. Diamond. 1975. Differential staining of the satellite regions of human acrocentric chromosomes. Experientia 31, 260-262.
Hsu, T.C., B.R. Brinkley and F.E. Arrighi. 1967. The structure and behavior of the nucleolus organizers in mammalian cells. Chromosoma 23, 137-153.
Hsu, T.C., S.E. Spirito and M.L. Pardue. 1975. Distribution of 18 + 28S ribosomal genes in mammalian genomes. Chromosoma 53, 25-36.
Hubbell, H.R., L.I. Rothblum and T.C. Hsu. 1979. Identification of a silver binding protein associated with the cytological silver staining of actively transcribing nucleolar regions. Cell Int. Rep. 3, 615-622.
Jordan, E.G. and C.A. Cullis. 1982. The Nucleolus. Cambridge: Cambridge University Press.
Lafontaine, J.G. 1968. Structural components of the nucleus in mitotic plant cells. In The nucleus, A.J. Dalton and Fr. Haguenau, ed., 151-196. New York: Academic Press.
Lepoint, A. and G. Goessens. 1982. Quantitative analysis of Ehrlich tumour cell nucleoli during interphase. Exp. Cell Res. 137, 456-459.
McClintock, B. 1934. The relation of a particular chromosomal element to the development of the nucleoli in Zea mays. Z.Zellforsch. 21, 294-328.
Mirre, C. and A. Stahl. 1978. Peripheral RNA synthesis of fibrillar center in nucleoli of Japanese quail oocytes and somatic cells. J. Ultrastr. Res. 64, 377-387.
Ochs, R.L. and H. Busch. 1984. Further evidence that phosphoprotein C23 (110 Kd/pI 5.1) is the nucleolar silver staining protein. Exp. Cell Res. 152, 260-265.
Ploton, D., M. Ménager and J.J. Adnet. 1985. Simultaneous ultrastructural localization of Ag-NOR (nucleolar organizer region) proteins during mitosis, in human breast cancerous tissues. J. Cell Sci. 74, 239-256.
Ploton, D., M. Thiry, M. Ménager, A. Lepoint, J.J. Adnet and G. Goessens. 1986. Behaviour of nucleolus during mitosis. A comparative ultrastructural study of various cancerous cell lines using Ag-NOR staining procedure. Chromosoma in press.
Puvion, E. and G. Moyne. 1981. In situ localization of RNA structures. In The cell nucleus, vol. 8, H. Busch, ed., 59-115. New York: Academic Press.

Ritossa, F.M. and S. Spiegelman. 1965. Localization fo DNA complementary to ribosomal RNA in the nucleolus organizer region of Drosophila melanogaster. Proc. Natl. Acad. Sci. USA. 53, 737-745.
Scheer, U. and K.M. Rose. 1984. Localization of RNA polymerase I in interphase cells and mitotic chromosomes by light and electron microscopic immunocytochemistry. Proc. Natl. Acad. Sci. USA. 81, 1431-1435.
Schwarzacher, H.G. and F. Wachtler. 1983. Nucleolus organizer regions and nucleoli. Human Genet. 63, 89-99.
Smetana, K. and H. Busch. 1974. The nucleolus and nucleolar DNA. In The cell nucleus, vol. 1, H. Busch, ed., 73-147. New York: Academic Press.
Smetana, K., R. Ochs, M.A. Lischwe, F. Gyorkey, E. Freireich, V. Chudomel and H. Busch. 1984. Immunofluorescence studies on proteins B23 and C23 in nucleoli of human lymphocytes. Exp. Cell Res. 152, 195-203.
Stahl, A. 1982. The nucleolus and nucleolar chromosomes. In The Nucleolus, E.G. Jordan and C.A. Cullis, ed., 1-24. Cambridge: Cambridge University Press.
Thiry, M. 1986. Ultrastructural study of nucleolar organazing regions by means of serial sections in Ehrlich tumour cells. Bull. Soc. Roy. Sci. Liège 55, 527-538.
Thiry, M. and G. Goessens. 1986. Ultrastructural study of the relationships between the various nucleolar components in Ehrlich tumour and HEp-2 cell nucleoli after acetylation. Exp. Cell Res. 164, 232-242.
Thiry, M., A. Lepoint and G. Goessens. 1985. Re-evaluation of the site of transcription in Ehrlich tumour cell nucleoli. Biol. Cell 54, 57-64.
Wallace, H. and M.L. Birnsteil. 1966. Ribosomal cistrons and the nucleolar organizer. Biochem. Biophys. Acta 114, 296-310.
Williams, M.A., J.A. Kleinschmidt, G. Krohne and W.W. Franke. 1982. Argytophilic nuclear and nucleolar proteins of Xenopus laevis oocytes identified by gel electrophoresis. Exp. Cell Res. 137, 341-351.

Nucleolar organisers in plants

E. G. Jordan

Department of Biophysics, Cell and Molecular Biology, King's College London, The Strand, London WC2R 5RL

INTRODUCTION

The nucleolar organiser region of plant chromosomes was identified by Heitz in 1931. He showed the correlation between size and number of secondary constrictions at metaphase and the size and number of nucleoli arising in telophase in a range of *Vicia* species. McClintock (1934) however working with a translocation line in *Zea* argued against the hypothesis of Heitz, coining the term 'nucleolar organising body' to describe a large knob of heterochromatin adjacent to the secondary constriction. Although McClintock gave us the concept of 'nucleolar organizing', the capacity to collect or 'organise' the scattered precursors of the nucleolus, she mistakenly attributed this function to the heterochromatic knob. The explanation for this apparent mistake has now become clear through the work of Phillips and his colleagues (1978) who have shown that rRNA precursor genes (ie the organising region by current definitions) are present in the heterochromatic knob and also in the secondary constrictions but that the heterochromatic knob is usually a 'silent' or non-transcribing region. The ribosomal genes are not all therefore regulated in the same way. The secondary constrictions are parts of the organisers which have incompletely condensed chromatin complexed with 'organiser' proteins. It is these regions that were transcribing in the preceding interphase that will be active again in nucleolar organising in the following interphase. It is merely the reduced staining levels consequent upon the dispersion of the chromatin and its association with protein that give the appearence of a constriction at the optical microscope.

In this paper attention is directed to the interphase arrangement of the segments of nucleolar organisers of plants, especially the alga Spirogyra. The interphase nucleolus has many pale staining regions - the fibrillar centres. (Terms are used according to the Banyuls' Nomenclature Agreement (Jordan,1984).) These have been identified as the secondary constrictions at interphase by the demonstration that they have the same relationships of size and number to nucleoli that Heitz showed for metaphase secondary constrictions (Jordan et.al.,1982).

Although there is general agreement that fibrillar centres are nucleolar organiser regions this must still be regarded as a hypothesis for at certain times the strict equivalence may not hold. Examples exist of nucleoli that do not normally show fibrillar centres and it is also reasonable to ask whether under some circumstances fibrillar centres might be devoid of chromatin. The chromatin component of the fibrillar centres of wheat nucleoli has been calculated to be of the order of 50% (Jordan et.al., 1982). If this chromatin component were to be withdrawn would the fibrillar centres look any different? Fibrillar centres are the first component to appear in nucleologenesis in activated avian erythrocytes (Hernandez-Verdun and Bouteille, 1979) and it could easily be envisaged that they appear before they are invaded by the chromatin carrying the rRNA genes. A parallel may be drawn with another component of nucleoli that normally contains chromatin, namely the dense fibrillar component. This component persists at times when transcription is arrested or when no active transcripts are present, yet it is usually identified as the component of nucleoli that contains the nucleolar organiser in the form of actively transcribing units or 'Christmas trees'. There remains therefore a need to fully clarify the nature of the fibrillar centres of nucleoli.

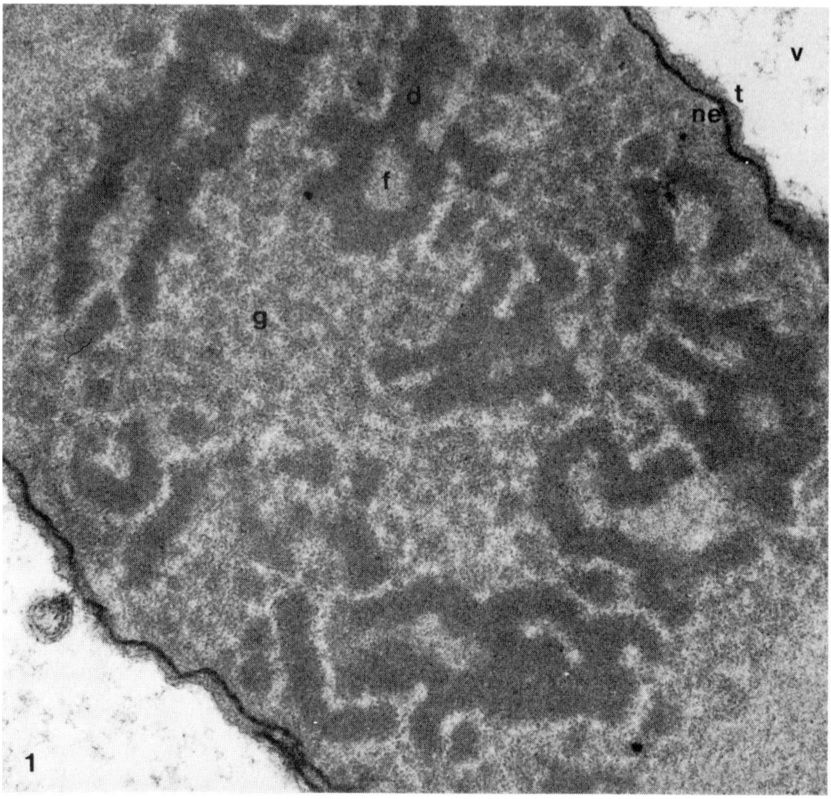

Fig 1. Nucleolus of Spirogyra showing fibrillar centres f, dense fibrillar component d, and granular component g. ne, nuclear envelope, t tonoplast, v, cell vacuole. x 20 000.

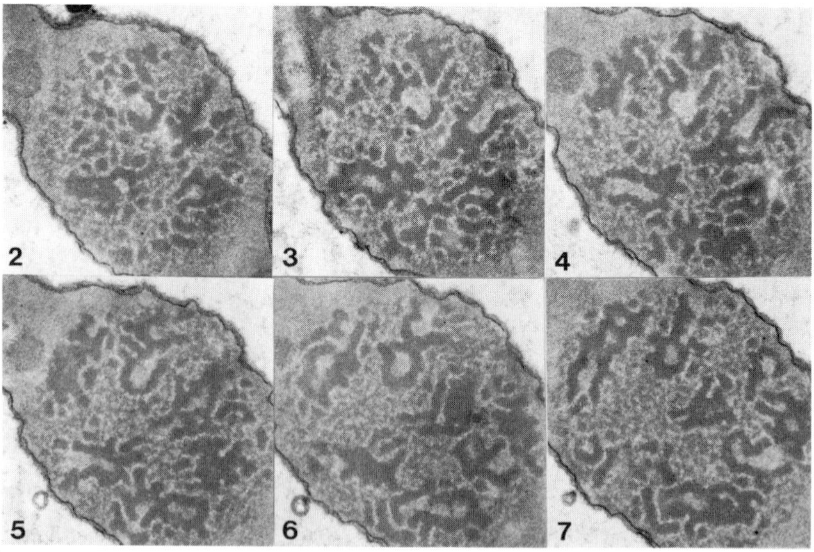

Figs 2-7. Adjacent serial sections showing the continuity of fibrillar centres. x 6 500.

CONTINUITY OF FIBRILLAR CENTRES

In most plant and animal nucleoli the fibrillar centres are scattered throughout the nucleolus though they always lie next to the dense fibrillar component. It has been the basis of some models (Stahl 1982) that these fibrillar centres are all connected by virtue of their association with the nucleolus organising region of the chromosome. This continuity was best demonstrated by Godward in Spirogyra in 1950 but it rested on optical microscope identification. LaCour (1966) isolated continuous structures from plant nucleoli and the demonstration of necklaces of beads in animal nuclei also confirms this view. In this paper, complete serial sectioning of Spirogyra nucleoli at the level of the electron microscope is used to demonstrate this continuity of fibrillar centres. Fig.1 shows the arrangement of fibrillar centres in a Spirogyra nucleolus. The profiles are not only circular as in so many animal cells but show the outlines obtained from sectioning through an elongated object. Figs. 2 to 7 are adjacent sections demonstrating a continuity of elongated structures. In Figs. 8 and 9 the superimposed outlines from all the fibrillar centres from two nucleoli are shown. They present such a mass of coiled structures that their overall number and distribution cannot be visualised, but they confirm the earlier description of the path of the organiser in Spirogyra shown by Godward in 1950. Figs. 10 to 13 show this same procedure applied separately to four parts of one nucleolus. Figs. 2 to 7 show how much the structure changes between adjacent 0.1μm sections. The continuity of the fibrillar centres is usually fairly clear but in a few places it seems that they disappear and their continuation is only suggested by the arrangement of the dense fibrillar component. There are also places where several ends seem to meet in a larger fibrillar centre and the relationship of these to continuity is unclear. For these reasons it is not possible to be certain of the precise number of separate loops of fibrillar centre in the nucleoli so far examined but it is between 6 and 10.

Although earlier studies made on prophase nucleoli have established that the fibrillar centres are connected to other

8

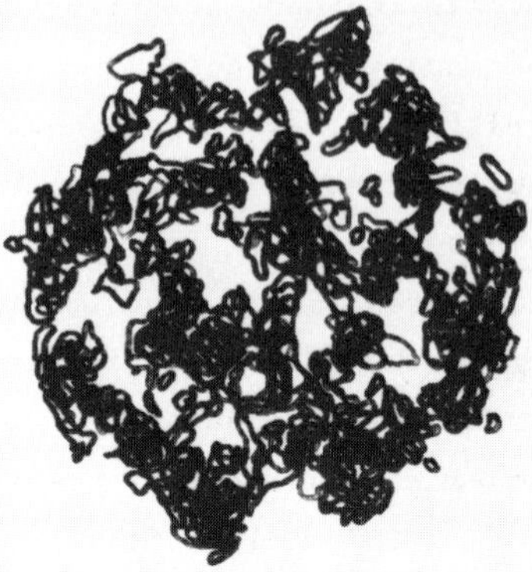

9

Figs 8 and 9. The superimposed outlines of all the fibrillar centres for two different Spirogyra nucleoli.

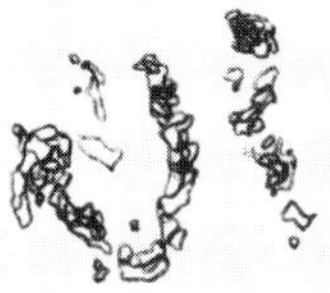

10

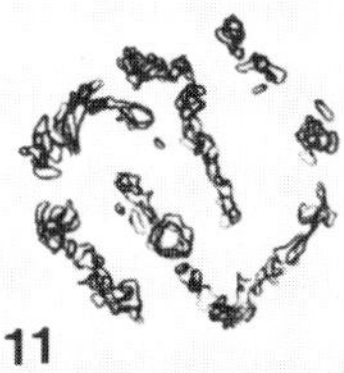

11

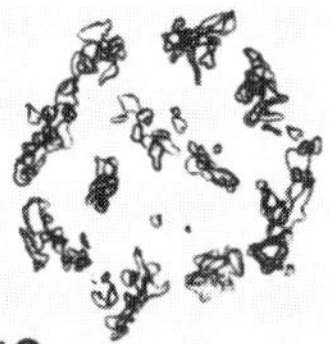

12

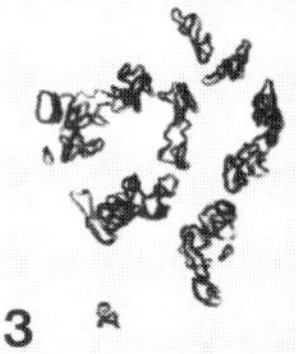

13

Figs 10-13. The tracings of Fig. 9 divided into 4 consecutive sets.

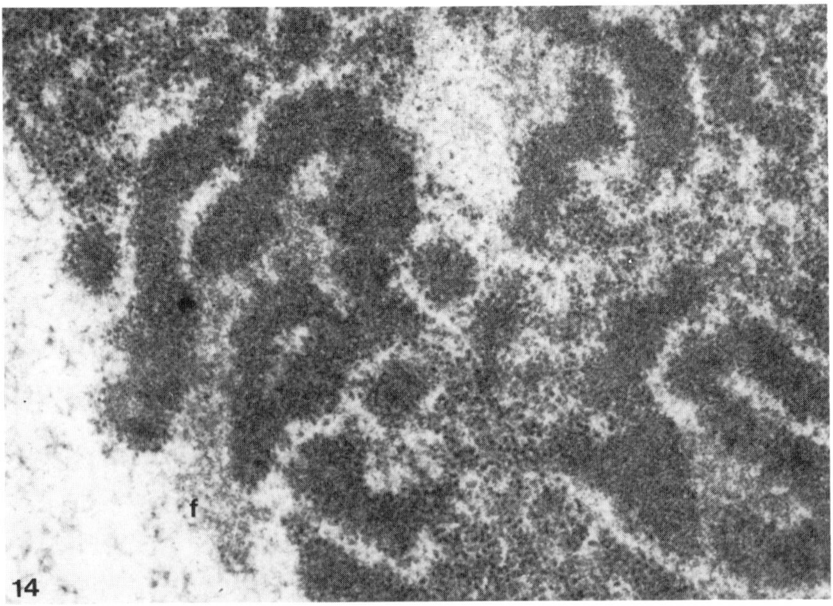

14

Fig. 14. The fibrillar centre, f, at the junction with the nucleoplasm. x 60 000.

chromosome regions, in this study of interphase nucleoli the connection to non-nucleolar chromatin could not be shown. It is interesting in view of this that earlier work reported chromosomes in some species of Spirogyra consisting entirely of nucleolar organising regions. It would seem unlikely that all eight or so fibrillar centres in this species are whole chromosomes or that all the chromosomes were confined to the nucleolus, even though this is the situation in late prophase for all species. Nevertheless the fibrillar centres could be seen to come to the surface of the nucleolus in places, Fig. 14, where they were not enclosed by dense fibrillar component but apparently came into contact with the nucleoplasm.

THE STATE OF CHROMATIN IN THE FIBRILLAR CENTRES.

In earlier studies of prophase stages in Spirogyra (Godward and Jordan, 1965) the difference between fibrillar centres and other chromosome regions was shown not to be very marked unlike in higher plants. In Helianthus tuberosus the diameter of the fibrils has been shown to be half as thick in the fibrillar centres, 10nm as opposed to 20 or more in the condensed chromatin (Jordan and Chapman, 1971). The most striking difference is the lower density of staining in the fibrillar centres with conventional EM stains Figs. 15 to 18. Such junctions are best shown when the nucleolus is not active in transcription. Fig. 15 shows the intermingling of the two types of chromatin in Triticum aestivum where transcription has been arrested by elevating the temperature to 35°C. Fig. 16, from inactive cells of Daucus carota, shows the condensed chromosome region penetrating deeply into the nucleolus but joining it via the intervening material of the fibrillar centre. Fig. 17, from an inactive cell of Helianthus tuberosus, shows a junction where the fibrillar centre is at the surface of the nucleolus. In some plants the fibrillar centres have many smaller regions of condensed chromatin inside them, Fig. 18, the so-called heterogenous fibrillar centres.

Since transcription has never been demonstrated in the fibrillar centres of nucleoli we can describe them as both functional and structural intermediates between the fully

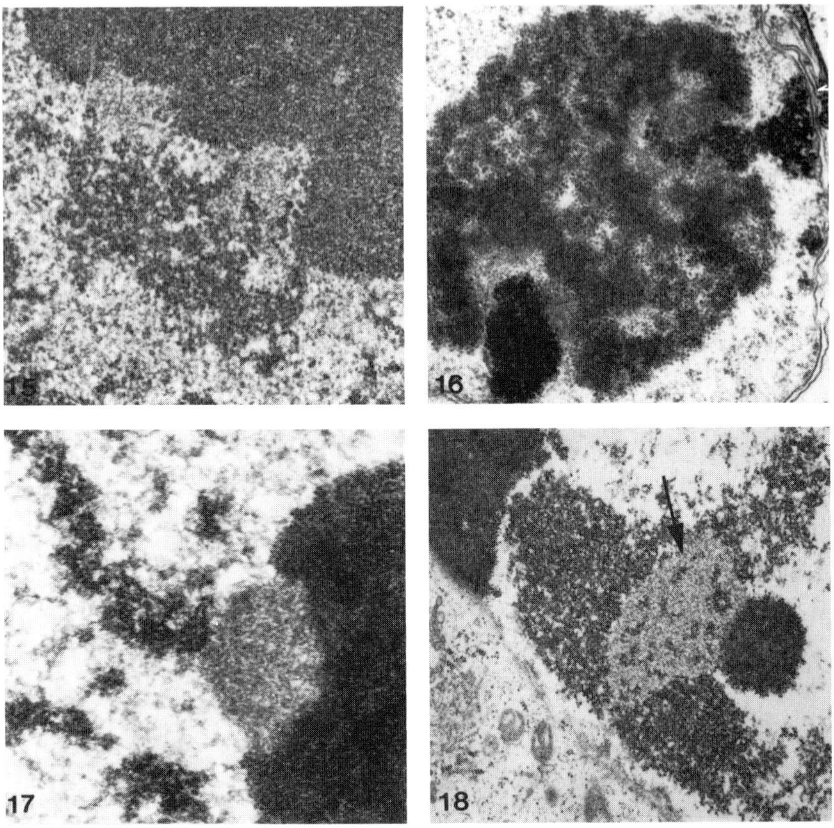

Fig. 15 Junction between fibrillar centres and condensed chromatin in Triticum aestivum root tip. x 32 000
Fig. 16 Condensed chromatin at junction with nucleolus showing intervening fibrillar centre material. Daucus carota parenchyma x 23 000
Fig. 17 Condensed chromatin joining nucleolus via fibrillar centre. Helianthus tuberosus parenchyma. x 42 000
Fig. 18 Large heterogenous fibrillar centre, arrowed, in Tradescantia virginiana pollen mother cell. x 15 000

condensed chromatin and the transcribing genes, regions which in some way are 'primed' or 'potentiated' for transcription but require a further factor or factors for full expression.

THE NUCLEOLONEMA AND THE FIBRILLAR CENTRES.

One of the questions that can be addressed in a structural study is the precise arrangement of fibrillar centres in relation to the transcribing genes in the nucleolus. Does the arrangement of the different components of nucleoli in any way direct our thinking about the types of regulation that might be involved? There are two features which seem to be significant. First the abrubt change at the boundary between the fibrillar centre and the dense fibrillar component of the nucleoli and secondly, the thick threads and sheets into which the dense fibrillar component is formed and the arrangement of these 'nucleolonemal' structures around the fibrillar centres. The threads of the nucleolonema can be thought of as being densly packed transcription units whose axis, the DNA, lies along the middle of the thread, the granules becoming apparent as the whole mass of transcripts moves away en bloc from the fibrillar centre Fig. 19. This would give rise to

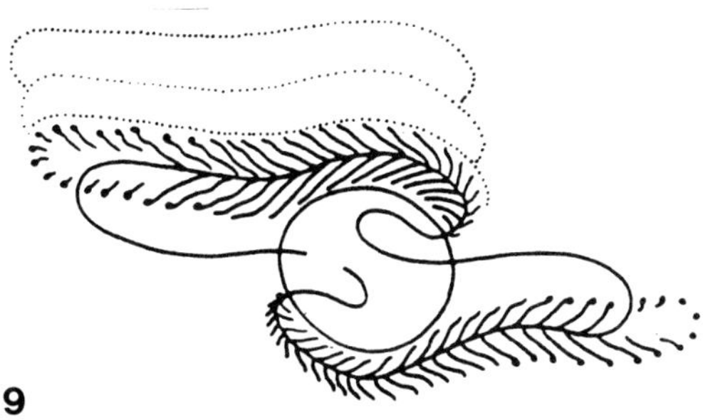

19

Fig. 19. Interpretation of arrangement of fibrillar centre and transciption units.

20

Fig. 20. Transcription at the surface of a fibrillar centre.

an interpretation of the nucleolonemal sheets as laterally opposed finger like protrusions from the fibrillar centres as shown in Fig. 19. The arrangement of the axes of transcription would conform to the lines in Fig. 21 but not Figs 22 or 23. The arrangement of the nucleolonemal sheets is sometimes suggestive of

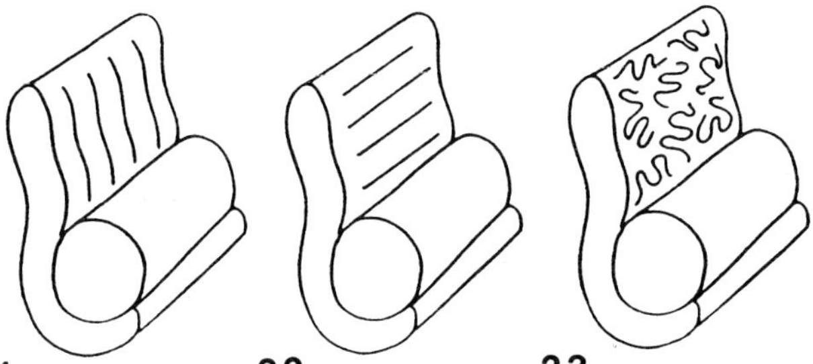

21 22 23

Fig 21, 22 and 23. Lines in the diagram show possible positions for the transciption units or their relics in the nucleolonema.

'waves' of transcription and processing radiating out evenly from the fibrillar centres. A model of this sort is shown in Fig. 24 but it seems less probable than the former. If the surface of the fibrillar centre is the site of transcription then all the transcripts would have to radiate out from it as shown in Fig. 20.

Fig. 24. This diagram shows how nucleolonemal sheets in concentric layers could arise as radiating 'waves' from the fibrillar centre.

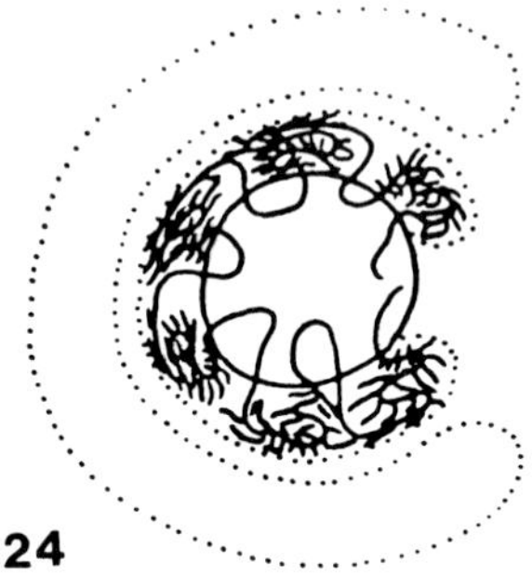

24

One of the major puzzles to be unravelled is the failure of conventional thin section electron microscopy to visualise granules in these early transcription structures when they are evident in spread preparations, presumably of the same component. The granules seen in spread preparations on the ends of the RNA transcripts may result from the process of spreading. Alternatively many of the transcription units that are seen in spread preparations could have been derived from the granular components of nucleoli. The latter explanation seems to be favoured by two observations. First, that the only structures which show any of the features expected of 'Christmas trees' in thin section are granular components. Secondly, where transcription units were unequivocally demonstrated by thin section study in the Balbiani rings of Chironomus they were seen in areas with strong similarity to nucleolar granular components (Lamb and Daneholt, 1979). Nevertheless autoradiography shows that the early stages of transcription are in the dense fibrillar component (See Goessens, this volume).

NUCLEOLONEMA WITHOUT DNA AXIS

In Spirogyra (Fig 25) the nucleolonemal arrangement of nucleoli is emphasised at prophase at the stage when fibrillar centres are withdrawn into the condensing chromosomes (Godward and Jordan 1965, Jordan and Godward 1969). All the granular component

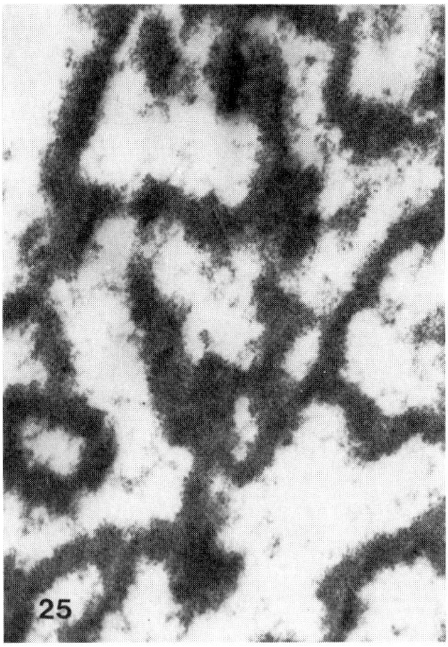

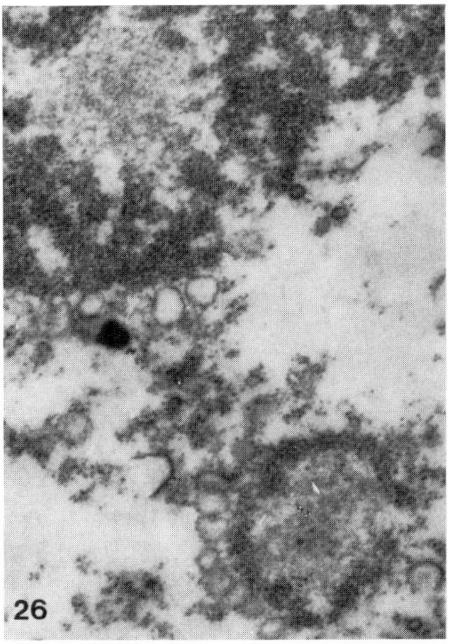

Fig. 25 Nucleolonema at prophase in Spirogyra after withdrawal of the fibrillar centres and loss of the granular component. x 28 000
Fig. 26 The persisting nucleolonema at anaphase in Spirogyra surrounding mitotic chromosomes, c. The nucleolonema is clearly granular. x 36 000

is lost at this stage and the persisting nucleolonemal structure is dense fibrillar component. By anaphase the structure of the nucleolonemal strands has become more granular, Fig. 26, suggesting that processing of transcripts continues. This is supported by the appearence of large accumulations of ribosomes around the telophase nuclei. A similar persistence and even augmentation of nucleolonemal morphology is apparent in prophase of Triticum Figs. 27 and 28. These nucleolonemal structures may have some supporting framework enabling them to persist in the absence of the DNA axis which would have been present during the transcription stages. The current ideas about nuclear skeleton and the attachment of RNA to such a skeleton at all times during its transcription, processing, translocation and translation (Jackson and Cook, 1985) would point to the fibrillar centres as central structures in such events. The skeletal fibres would presumably move with the transcripts through the cell or provide a

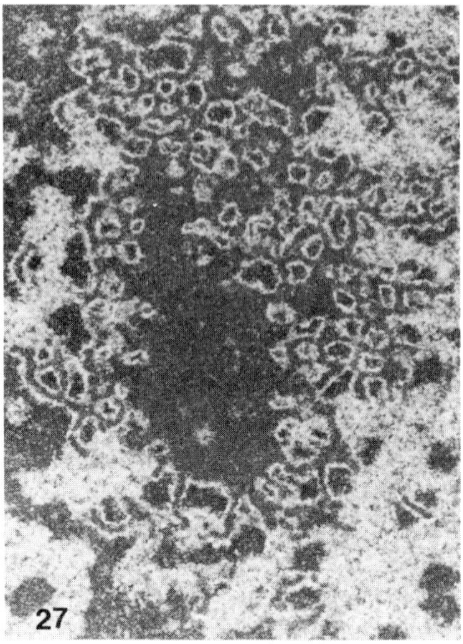

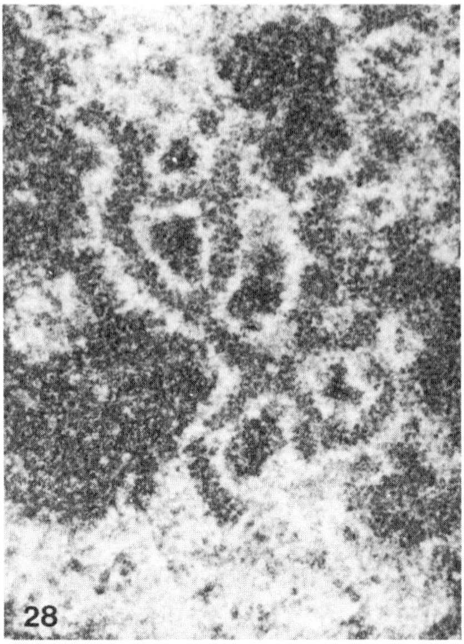

Fig. 27 Nucleolus at prophase in Triticum aestivum showing the accentuation of the nucleolonemal morphology. x 11 000
Fig. 28 Detail of Fig 22 showing the granular nature of the nucleolonemal sheets. x 38 000

framework along which they could migrate. The absence of visible fibrillar centres in some situations can also be explained by the spreading out of such a supporting protein network into single fibres. Transciptional arrest resulting in the appearence or enlargement of fibrillar centres could be the reverse of this dispersal.

These studies emphasise that the nucleolus with its several components and high multiplicity of genes can provide a vivid visual account of some of the steps in genetic transcription and processing and equip us with one system in which we might hope to learn more about the structural counterparts of the regulatory steps involved in gene expression. It should become one of our chief goals to fully clarify the in situ arrangement and subsequent history of the transcripts of nucleolar organisers.

REFERENCES

Godward, M.B.E. 1950. On the nucleolus and nucleolar organising chromosomes of Spirogyra. Ann. Bot. 14, 39-53.

Godward, M.B.E. and Jordan, E.G. 1965. Electron microscopy of the nucleolus of Spirogyra brittanica and Spirogyra ellipsospora. Jl.R. microsc. Soc. 84, 347-60.

Heitz, E. 1931. Nukleolen und chromosomen in der gattung Vicia. Planta 15, 495-505.

Hernandez-Verdun, D. and Bouteille, M. 1979. Nucleologenesis in chick erythrocyte nuclei reactivated by cell fusion. J. Ultrastruct. Res. 69, 164-79.

Jackson, P.R. and Cook, P.R. 1985. Transcription occurs at a nucleoskeleton. The EMBO Journal 4, 919-25.

Jordan. E.G. 1984. Nucleolar Nomenclature. J. Cell Sci. 67, 217-220.

Jordan. E.G. and Chapman, J.M. 1971. Ultrastructural changes in the nucleoli of Jerusalem artichoke (Helianthus tuberosus) tuber discs. J. Exp. Bot. 22, 627-34.

Jordan, E.G., Bennett, M.D. and Smith, J.B. 1982. Nucleolar organisers and fibrillar centres in Triticum aestivum L., c.v. Chinese Spring. Chromosoma 87, 447-59.

Jordan, E.G. and Godward, M.B.E. 1969. Some observation on the nucleolus in Spirogyra. J. Cell Sci. 4, 3-15.

La Cour, L.F. 1966. The internal structure of nucleoli. In Chromosomes Today vol. 1, C.D. Darlington and K.R. Lewis, eds., 150-160. Edinburgh and London: Oliver and Boyd.

Lamb, M.M. and Daneholt. B. 1979. Characterisation of active transcripton units in Balbiani rings of Chironomus tentans. Cell 17, 835-48.

McClintock. B. 1934. The relation of a particular chromosome element to the development of the nucleoli in Zea mays. Z. Zellforsch. mikrosk. Anat 21, 294-328.

Phillips. R. L. 1978. Molecular cytogenetics of the nucleolus organizer region. In Maize breeding and genetics. D.B. Walden, ed., 711-41. Wiley: New York.

Stahl, A. 1982. The Nucleolus and nucleolar chromosomes. In The Nucleolus. C.A. Cullis and E.G.Jordan Eds., 1-24. Cambridge University Press: Cambridge.

Immunocytochemical localization of RNA polymerase I in the fibrillar centers of nucleoli

U. Scheer[1] and I. Raska[2]

[1]Institute of Cell and Tumor Biology, German Cancer Research Center, D–6900 Heidelberg, Federal Republic of Germany, and [2]Institute of Experimental Medicine, Czechoslovakian Academy of Sciences, Prague, CSSR

For an understanding of the functional organization of the nucleolus identification of the site of the expression of the rRNA genes is of considerable importance. Although various stages of preribosome formation have been mapped to morphologically distinct components of the nucleolus (Hadjiolov, 1985), the intranucleolar location of transcriptionally active rRNA genes has remained uncertain. From high resolution autoradiographic studies, following pulse labelling of cells with radioactive RNA precursors, it was suggested that transcription of rRNA genes takes place in the "dense fibrillar component", and possibly also at the periphery of the "fibrillar centers" (Fakan, 1978; Thiry et al., 1985). However, using other techniques such as autoradiography with tritiated thymidine or actinomycin D, in situ-hybridization with radioactive rRNA probes and selective DNA-staining with the osmium-amine Feulgen-like reaction, DNA (including rDNA) could be detected only in the fibrillar centers (for reviews see Hernandez-Verdun, 1983; Goessens, 1984). To reconcile these seemingly contrasting results it was proposed that the fibrillar centers contain inactive rDNA in a relatively high packing density, thereby facilitating its detection, whereas the transcribed rRNA genes loop out into the surrounding dense fibrillar component (Stahl, 1982; Goessens, 1984). According to this concept the dense fibrillar component is formed by the superposition of rRNA-transcription units and newly formed rRNP material.

Recently, we took an alternative approach to locate transcriptionally active rRNA genes within the nucleolus which was based on immunocytochemical techniques using antibodies specific for RNA polymerase I (Scheer and Rose, 1984). The results indicated that active rRNA genes are specifically confined to the fibrillar centers of rat liver nucleoli. However, since the "pre-embedding" technique used in that study may result in artifacts of selective "masking", i.e. differential accessibility of the various nucleolar components to the antibodies, we have extended these studies to antibody labelling on ultrathin frozen sections of cultured mammalian cells.

IMMUNOFLUORESCENCE MICROSCOPY

Rabbit antibodies to RNA polymerase I (RP I) stain the nucleoli of cultured cells derived from various species (Figs. 1a, 2b,c and 3b; for details see Scheer and Rose, 1984). The nucleolar fluorescence of rat cells shown in Figure 1 appears as a punctate pattern, indicating that the RP I is dispersed in several distinct subnucleolar entities (Fig. 1a; nucleoli of this cell type contain a large number of small fibrillar centers as seen in the electron microscope). A different immunofluorescence pattern is observed in cells that posess only a few, but relatively large fibrillar centers such as mouse Ehrlich ascites cells as demonstrable by electron microscopy (Fig. 2a). When these cells are grown on coverslips or processed as frozen sections of cell pellets and examined by immunofluorescence microscopy using antibodies to RP I, the staining appears in a small number (usually 2-3) of relatively large, distinct intranucleolar entities (Fig. 2b,c). The homogeneous immunostaining suggests that the RP I-molecules are distributed throughout the fibrillar centers and are not confined to the peripheral layer of the dense fibrillar material.

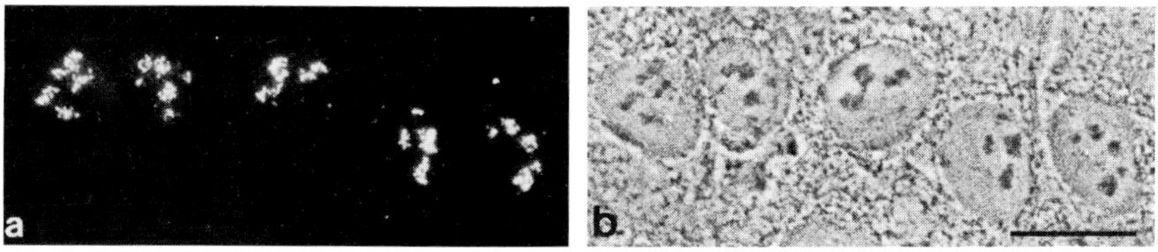

FIGURE 1. Immunofluorescence microscopy of cultured rat cells (RVF-SMC, for refs. see Scheer and Rose, 1984) after staining with antibodies to RPI. Note the numerous intranucleolar dot-like substructures (a). The corresponding phase-contrast micrograph is shown in b. Bar = 20 μm.

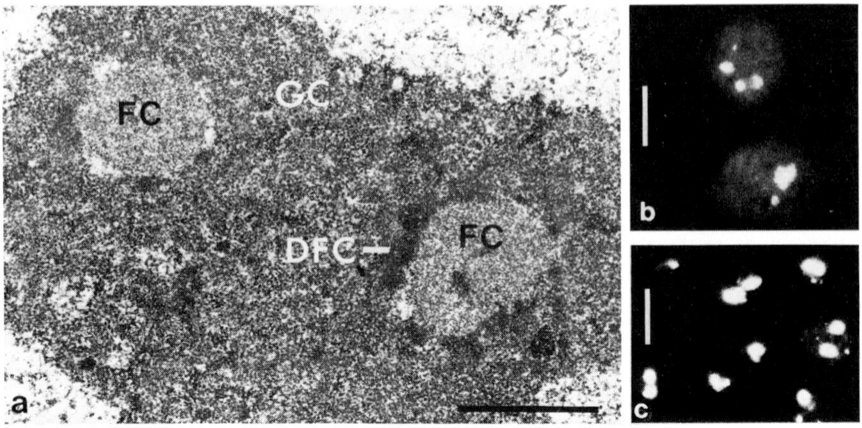

FIGURE 2. (a) Electron microscopy of a sectioned mouse cell (Ehrlich ascites) showing two large fibrillar centers (FC) embedded in the granular component of the nucleolus (GC). The dense fibrillar component (DFC) forms a thin layer around the fibrillar centers. (b,c) Immunofluorescence microscopy of Ehrlich cells grown on coverslips (b) and of a cryostat section through a cell pellet (c) after incubation with anti-RP I. The nucleoli contain large fluorescent spheres. Bars = 1 μm (a); 10 μm (b,c).

Nucleoli of cultured Xenopus cells also contain relatively large fibrillar centers (Fig. 3a). Correspondingly, immunofluorescence microscopy with antibodies to RP I reveals intranucleolar aggregates of distinctly fluorescing spheres (Fig. 3b).

ELECTRON MICROSCOPIC CHROMATIN SPREAD PREPARATIONS

Very brief exposure of cells to low salt buffer, prior to centrifugation onto electron microscopic grids, retain to some extent the original spatial arrangement of the rRNA genes, although in such preparations the topological organization of the various nucleolar components is usually lost. As shown in Figure 3d, distinct roundish clusters each containing 10 to 30 closely packed and intensely transcribed rRNA genes are observed when chromatin of cultured Xenopus cells is gently spread for electron microscopy.

ACTINOMYCIN D (AMD)-INDUCED REDISTRIBUTION OF RNA POLYMERASE I

Inhibition of transcription by exposure of Xenopus cells to AMD leads to a rearrangement of the fibrillar centers and the other nucleolar components ("nucleolar segregation"). Typically the fibrillar centers appear in a more peripheral location, partly surrounding the segregated dense fibrillar component (Fig. 3e). This rearrange-

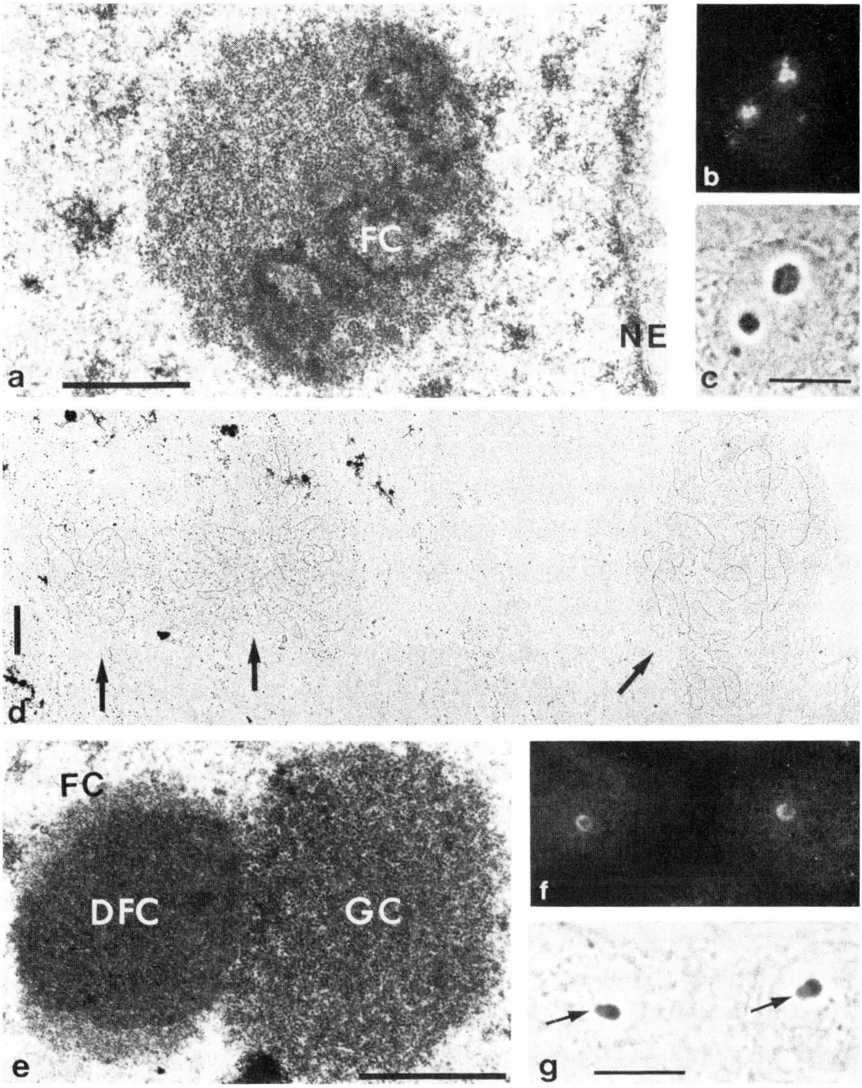

FIGURE 3. Cultured Xenopus laevis kidney epithelial cells (line A6; American Type Culture Collection). (a) Electron microscopy reveals the three nucleolar components (see Fig. 2). NE, nuclear envelope. (b,c) Immunofluorescence microscopy (c, phase contrast) after staining with anti-RP I. (d) Chromatin spread preparation after very brief lysis of Xenopus cells in low salt buffer (for technical details see Scheer and Zentgraf, 1982). Active rRNA genes occur in several compact aggregates (arrows). (e) Electron microscopy (ultrathin section) of an segregated nucleolus after 4h of AMD-treatment (5 μg/ml). (f,g) Immunofluorescence microscopy of AMD-treated Xenopus cells after staining with anti-RP I. The fluorescent cap (f) is situated on top of the phase-light dense fibrillar component (arrows in g). Bars = 1 μm (a,d,e); 10 μm (c,g).

ment of the nucleolar components upon AMD-treatment results in a considerable decrease in immunostaining intensity and a redistribution of the RP I-containing structures from an intranucleolar (Fig. 3b) to a peripheral (Fig. 3f) position. The crescent-shaped RP I-positive component of segregated nucleoli always surrounds the dense fibrillar component (Fig. 3f,g).

IMMUNOGOLD ELECTRON MICROSCOPY

The "pre-embedding" labelling technique as described previously (Scheer and Rose, 1984) has the advantage that the antibody binding reaction can be performed on frozen tissue sections (about 5 µm thick) without any prior chemical fixation. For example, in Figure 4 individual nucleolar components of a rat hepatocyte are clearly recognized. Gold particles, reflecting the distribution of the bound RP I antibodies, are enriched over the fibrillar centers and are clearly absent from the dense fibrillar component.

Immunolabelling of ultrathin frozen sections ("post-embedding" technique) gives essentially identical results. Figure 5 shows an ultrathin cryosection of fixed (4% paraformaldehyde) cultured PC12 rat cells incubated with RP I antibodies, followed by secondary

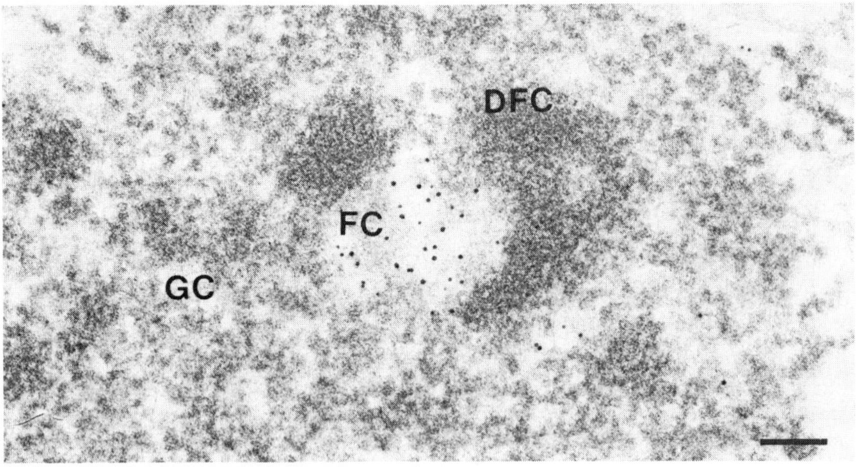

FIGURE 4. Electron microscopic immunolocalization of RP I in rat hepatocytes. Cryostat sections were incubated with antibodies to RP I followed by secondary antibodies coupled to 5 nm gold particles. After the immunoreaction the tissue section was processed for electron microscopy (see Scheer and Rose, 1984). Gold particles are selectively enriched over the fibrillar center (FC). Bar = 0.1 µm.

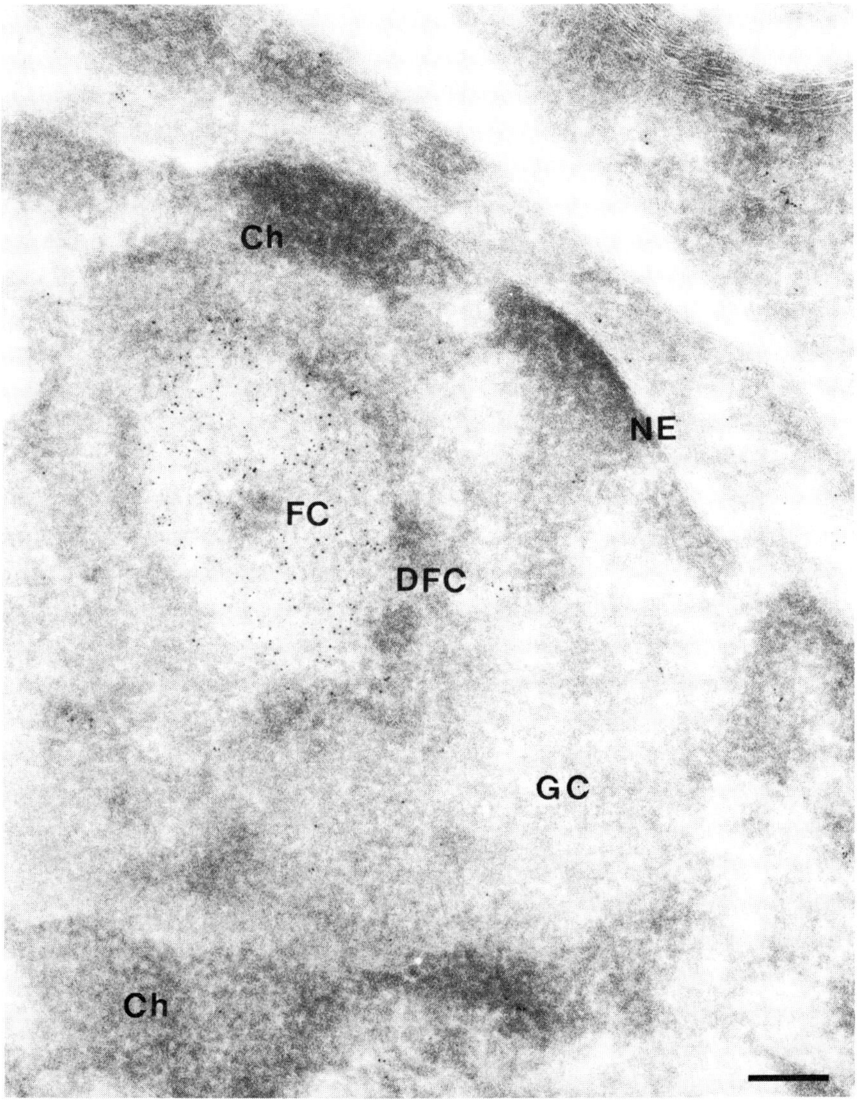

FIGURE 5. RP I antibody labelling of an ultrathin frozen section of cultured rat PC12 cells (Greene and Tischler, 1976). A cell pellet was fixed in 4% formaldehyde for 60 min, washed in PBS, incubated in 2.1 M sucrose and shock-frozen in liquid nitrogen. Cryosections were cut and incubated with antibodies essentially as described by Tokuyasu (1980). Secondary antibodies coupled to 5 nm gold particles were purchased from Janssen Life Sciences (Beerse, Belgium) and used in a 1:10 dilution. Gold particles are enriched over the large fibrillar center (FC) and are almost absent from the surrounding dense fibrillar component (DFC) and the granular component (GC). Ch, nucleolus-associated heterochromatin; NE, nuclear envelope. Bar = 0.2 μm.

antibodies coupled to collodial gold. The gold particles are selectively enriched over the large fibrillar center. All other nucleolar structures are not significantly labelled above background. Human autoantibodies to RP I, which occur in sera of scleroderma patients, also decorate specifically the fibrillar centers of these cells (Fig. 6; for details see Reimer et al., 1986). Again, there is no significant enrichment of gold particles in the dense fibrillar component.

CONCLUSIONS

The present results confirm and extend our previous conclusion based on a "pre-embedding" immunocytochemical approach that RP I is found exclusively in the fibrillar centers of nucleoli (Scheer and Rose, 1984). Does this result allow the conclusion that transcription of rDNA primarily or even exclusively takes place in this nucleolar compartment? It could be argued, for instance, that the binding of RP I antibodies to the fibrillar centers merely reflects the presence of a pool of "stored" enzymes bound to an unknown structure and that the transcribing polymerases are in fact present in the dense fibrillar component but not accessible to the antibodies. Moreover, the antibodies could recognize only a specific posttranslational modification present in inactive RP I but absent in transcribing molecules. However, several arguments speak against this view which can be summarized as follows.

1. <u>The absence of antibody labelling in the dense fibrillar component most likely reflects the absence of RNA polymerase I</u>. Antibody labelling on ultrathin frozen sections provides optimal accessibility of epitopes. Thus, most likely, the absence of gold label over the dense fibrillar component is not due to a selective "masking" effect. Another possibility, i.e. selective recognition of free vs. template-engaged RP I also seems unlikely since the rabbit antibodies used contain IgGs directed against epitopes of most of the RP I-subunits (Rose et al., 1981).

2. <u>Antibody binding to the fibrillar centers is not due to a storage pool of polymerases</u>. It is well known that RP I as well as the other RNA polymerase classes exist in two functionally distinct states, i.e. template-bound and free (for reviews see Grummt, 1978; Muramatsu et al., 1979). The free form of RP I is readily diffusible

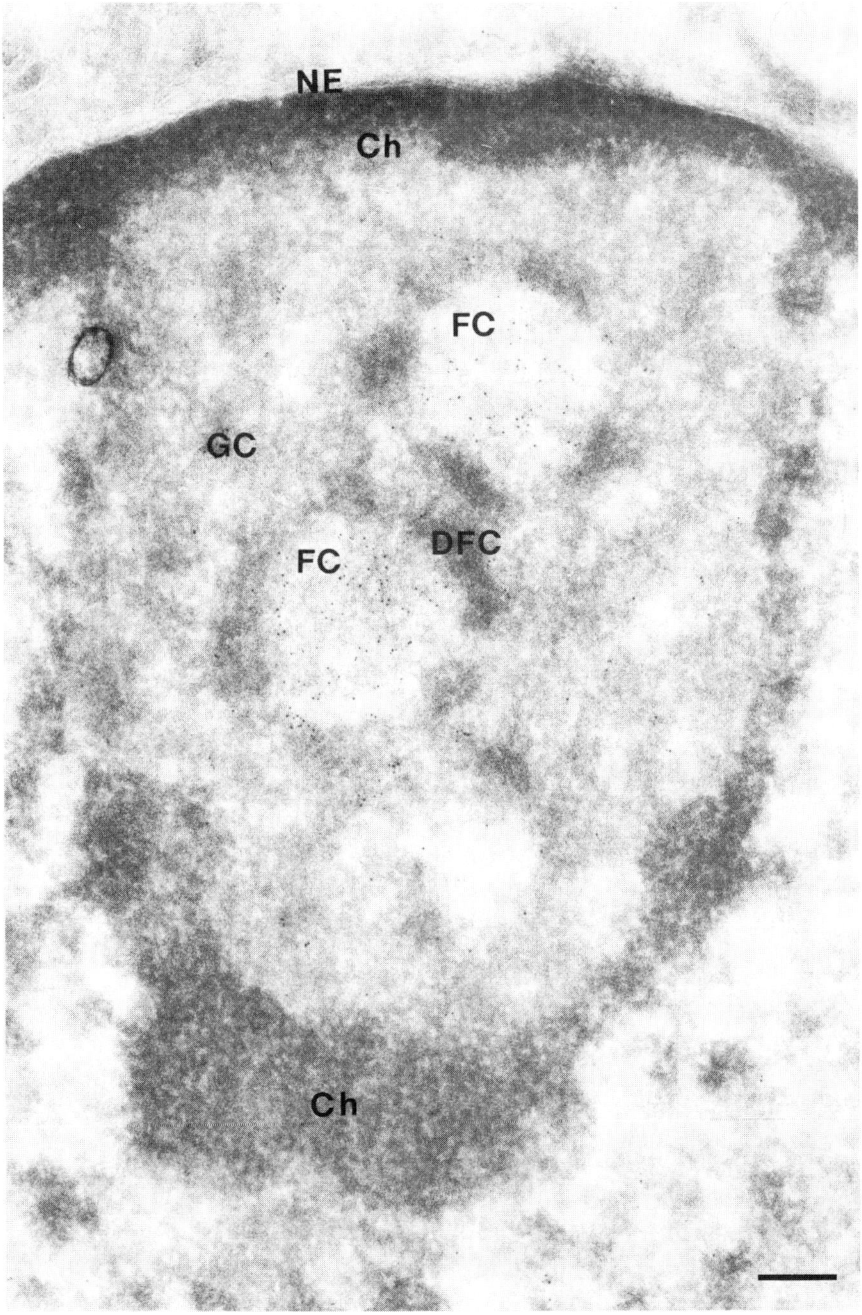

FIGURE 6. Similar preparation as shown in Fig. 5 but after incubation with an autoimmune serum from a scleroderma patient with anti-RP I specificity (Reimer et al., 1986). Again, gold particles are concentrated in the fibrillar centers (FC). Bar = 0.2 µm.

and released from isolated nuclei upon incubation in isotonic buffers (about 20% of rat liver RP I exists in the free form; Yu, 1980; see also Jacob and Rose, 1978). Furthermore, the observation that addition of exogenous DNA templates to nucleoli isolated under conditions minimizing RP I leakage stimulates RNA synthesis, also argues against a stable association of the free RP I to nucleolar structures (see Grummt, 1978; Muramatsu et al., 1979). That the soluble pool of RP I does not contribute to the present immunolocalization results is demonstrated by a considerable reduction in nucleolar immunostaining following treatment of cells with AMD. Short-term exposure of cells to AMD does not change the total amount of RP I-molecules present but shifts the equilibrium between free and engaged enzymes toward the free form.

3. Initiated but blocked polymerases are not detectable in nucleoli. This "frozen" status of RNA polymerases is detected when transcription of nuclei, chromatin or isolated chromosomes is assayed *in vitro* under conditions excluding new initiation events ("run-on" or "elongation"-assay). Stimulation of transcription by removal of DNA-associated proteins with the polyanion heparin or the anionic detergent Sarkosyl indicates the presence of pre-existing but blocked transcriptional complexes. When isolated nuclei were assayed with or without heparin, no significant stimulation of RP I-dependent transcription was observed (Coupar and Chesterton, 1977). Thus, it can be excluded that the polymerases present in the fibrillar centers represent a special form of initiated but transcriptionally inactive enzymes. A different situation is encountered during mitosis. RP I remains bound to the nucleolus organizing regions (NOR) of mitotic chromosomes (Scheer and Rose, 1984), apparently as initiated complexes with rDNA which, however, do not elongate (Matsui and Sandberg, 1985).

Taken together, we feel that our immunocytochemical data justify the conclusion that transcription of rRNA genes takes place in the fibrillar centers of nucleoli. As mentioned in the Introduction, this conclusion agrees with several studies showing that DNA including rDNA is present in the fibrillar centers but is at variance with certain autoradiographic studies of RNA labelling. However, it has to be kept in mind that pulse-labelling of cells with radioactive uridine necessarily leads to two different states of radioactive RNAs, i.e. template-attached and released molecules. Furthermore, a labelling period of 2-5 min which is often used (Fakan, 1978; Stahl, 1982; Goessens, 1984; Thiry et al., 1985) is about the time needed for an individual RP I molecule to transcribe an entire rRNA gene (approximately 3-5 min). Immediate translocation of the completed pre-rRNAs to and accumulation in the dense fibrillar component would lead to a high local packing density of RNA in this compartment. Thus, the distribution of silver grains might reflect primarily the site where pre-rRNAs accumulate after their release from the template rather than the site of transcription.

We trust that the precise location of transcribing rRNA genes within the nucleolus will enable us, by using a series of antibodies to different ribosomal and nucleolus-specific proteins, to understand in more detail the spatial and temporal patterns of interaction between pre-rRNAs and proteins involved in the formation of nucleolar preribosomal particles.

Acknowledgements: We thank Drs. W.W. Franke and F. Longo for their critical reading of the manuscript. We are grateful to Dr. K. Rose (University of Texas, Houston, Texas) for rabbit antibodies to RP I, Drs. G. Reimer and E. Tan (Scripps Clinic, La Jolla, California) for human autoimmune sera, and Dr. C. Granzow (this institution) for Ehrlich ascites cells. This work was supported by grants from the Deutsche Forschungsgemeinschaft (Sche 157/5-4).

REFERENCES

Coupar, B.E.H. and C.J. Chesterton. 1977. The mechanism by which heparin stimulates transcription in isolated rat liver nuclei. Eur. J. Biochem. 79, 525-533.

Fakan, S. 1979. High resolution autoradiography studies on chromatin functions. In The cell nucleus, vol. 5, H. Busch, ed., 3-53. New York: Academic Press.

Goessens, G. 1984. Nucleolar structure. Int. Rev. Cytol. 87, 107-158.

Greene, L.A. and A.S. Tischler. 1976. Establishment of a noradrenergic clonal line of rat adrenal pheochromocytoma cells which respond to nerve growth factor. Proc. Natl. Acad. Sci. USA 73, 2424-2428.

Grummt, I. 1978. In vitro synthesis of pre-rRNA in isolated nucleoli. In The cell nucleus, vol. 5, H. Busch, ed., 313-414. New York: Academic Press.

Hadjiolov, A.A. 1985. The nucleolus and ribosome biogenesis. Vienna: Springer-Verlag.

Hernandez-Verdun, D. 1983. The nucleolar organizer region. Biol. Cell. 49, 191-202.

Jacob, S.T. and K.M. Rose. 1978. RNA polymerases and poly(A) polymerase from neoplastic tissues and cells. In *Methods in cancer research*, vol. 14, H. Busch, ed., 191-241. New York: Academic Press.

Matsui, S. and A.A. Sandberg. 1985. Intranuclear compartmentalization of DNA-dependent RNA polymerases: association of RNA polymerase I with nucleolar organizing chromosomes. *Chromosoma* 92, 1-6.

Muramatsu, M., T. Matsui, T. Onishi and Y. Mishima. 1979. Nucleolar RNA polymerase and transcription of nucleolar chromatin. In *The cell nucleus*, vol. 7, H. Busch, ed., 123-161. New York: Academic Press.

Reimer, G., K.M. Rose, U. Scheer and E. Tan. 1986. Autoantibody to RNA polymerase I in scleroderma sera. Submitted to *J. clin. Invest*.

Rose, K.M., D.A. Stetler and S.T. Jacob. 1981. Protein kinase activity of RNA polymerase I purified from a rat hepatoma: probable function of M_r 42,000 and 24,600 polypeptides. *Proc. Natl. Acad. Sci. USA* 78, 2833-2837.

Scheer, U. and H. Zentgraf. 1982. Morphology of nucleolar chromatin in electron microscopic spread preparations. In *The cell nucleus*, vol. 11, H. Busch, ed., 143-176. New York: Academic Press.

Scheer, U. and K.M. Rose. 1984. Localization of RNA polymerase I in interphase cells and mitotic chromosomes by light and electron microscopic immunocytochemistry. *Proc. Natl. Acad. Sci. USA* 81, 1431-1435.

Stahl, A. 1982. The nucleolus and nucleolar chromosomes. In *The nucleolus*, E.G. Jordan and C.A. Cullis, eds., 1-24. Cambridge: Cambridge University Press.

Thiry, M., A. Lepoint and G. Goessens. 1985. Re-evaluation of the site of transcription in Ehrlich tumour cell nucleoli. *Biol. Cell.* 54, 57-64.

Tokuyasu, K.T. 1980. Immunocytochemistry on ultrathin frozen sections. *Histochem. J.* 12, 381-403.

Yu, F. (1980) High concentration of RNA polymerase I is responsible for the high rate of nucleolar transcription. *Biochem. J.* 188, 381-385.

Distribution of rDNA and of its transcription sites in the nucleolus of the human Sertoli cell

M. Devictor, M. Hartung and A. Stahl, with the technical assistance of A. de Lanversin

Laboratoire de Génétique, Faculté de Médecine, 27 boulevard Jean Moulin, 13385 Marseille Cédex 5, France

SUMMARY

The transcription sites of ribosomal genes were studied on autoradiographs following ^{3}H-uridine incorporation in human Sertoli cells. Transcription was localized over all dense fibrillar regions of the nucleolus, around and outside fibrillar centres. The location of transcription sites was compared with the distribution of ribosomal genes revealed by in situ hybridization. The results lead to the conclusion that rDNA uncoils outside the NOR represented by the fibrillar centre.

INTRODUCTION

The presence of one or two large fibrillar centres spontaneously separated from the other nucleolar components is a characteristic feature of the human Sertoli cell nucleolus (Devictor et al., 1984). By the distribution of its components, this nucleolus can be considered as a model for studying the functional significance of the fibrillar centre and the location of ribosomal genes. A large number of data suggest that the fibrillar centre is the interphasic counterpart of the NOR (Hernandez-Verdun et al., 1980 ; review in Stahl, 1982 and Goessens 1984). It has also been argued that the fibrillar centre might be mainly a storage site for proteins involved in rDNA transcriptional processes and rRNA maturation (Mirre and Knibiehler, 1982 ; Wachtler et al., 1980 and Schwarzacher and Wachtler, this volume). RNA polymerase I is localized within the fibrillar centre (Scheer and Rose, 1984 and Scheer and Raska, this volume). It has been shown in the mouse oocyte that rDNA, initially compacted within the fibrillar centre, uncoils and extends into the fibrillar strands of the growing nucleolus (Stahl et al., 1980 ; Mirre and Stahl, 1981).

In this study we show that the human Sertoli cell is another example of nucleolar organization where rDNA is distributed at the same time at the periphery of the fibrillar centre and within the fibrillo-granular strands that extend far from the fibrillar centre.

MATERIAL AND METHODS

The testicular samples originated from five normally fertile patients who were subjected to testicular surgical operation under general anesthesia.

EM autoradiography

Small pieces of testis were incubated for 20 min or 40 min at 25°C in 1 ml of 80 % Eagle's minimal medium 20 % calf serum containing ^{3}H-uridine at 100 µCi/ml (specific activity 24 Ci/mmol, CEA, France). After fixation and embedding, the sections were transferred onto a collodion-coated slide and coated by using an Ilford L4 emulsion diluted 1:4. After exposure for 2 months, the preparations were developed in Microdol-X (Kodak), fixed using Kodak rapid fixer and stained with uranyl acetate and lead citrate.

In situ hybridization

Sertoli cell preparations were obtained from the seminiferous tubules using the air-drying technique of Luciani et al.(1974). r RNA (28S) was isolated using a sucrose gradient from human thyroid polysomal RNA free of poly A+ RNA after chromatography on oligo-dT cellulose (T3 Collaborative Research). 28S ^{3}H cDNA (specific activity 0.9 to 1.1 x 10^{7} dpm/µg) was synthesized as described by Brahic, Stroop and Baringer (1981) using calf thymus DNA random primer purified by chromatography on DEAE cellulose and ^{3}H dCTP (50 to 80 Ci/mmol Amersham U.K.). The cell spreads were prepared for in situ hybridization according to the method of Gall and Pardue (1971), except that the DNA denaturation step was performed in 95 % formamide, 0.1 x standard saline citrate (NaCl/Cit) for 2 hr at 72°C. cDNA (28S) was hybridized to testicular cells in 50 % formamide, 3xNaCl/Cit, for 24 hr at 42°C in a sealed moist chamber. Twenty ng of cDNA in a 20 µl mixture covered by an 18 x18 acid washed coverslip were used per slide. Autoradiographs were prepared according to the method of Gall and Pardue (1971).

RESULTS AND DISCUSSION

At the ultrastructural level,the Sertoli cell nucleus was easily identified by the absence of dense chromatin and the presence of deep invaginations of the nuclear envelope extending sometimes close to the nucleolus.

The nucleolus displayed several characteristics : 1. The nucleolar components were separated, constituted by : a) one or two large fibrillar centres partially surrounded by strands of dense fibrils ; b) fibrillar or fibrillo-granular strands continuous with those surrounding the fibrillar centre ; c) a large granular part, often incompletely surrounded by fibrillo-granular strands. 2. Associated or perinucleolar chromatin was never observed.

On autoradiographs after ^{3}H-uridine incorporation for 20 min, the nucleoplasm was heavily labelled. Labelling intensity always exceeded that of germinal cell nuclei. The nucleolar labelling was localized only over the dense fibrillar component and the fibrillo-granular strands. Indeed, silver grains were observed around the fibrillar centre, between the fibrillar centre and the granular part and over the fibrillar and fibrillo-granular strands that extended along the granular part or at some distance from the latter (Fig. 1-3).

On the air-dried preparations used for <u>in situ</u> hybridization, the Sertoli cell nuclei were identified by their clear aspect and their single, centrally located and large nucleolus. When cells were stained with Giemsa, the site of the fibrillar centre was often recognizable by the presence of a pink-stained spherule.

Autoradiographs following <u>in situ</u> hybridization showed several patterns of distribution of labelling grains. In some nuclei, most grains were concentrated over a lateral zone corresponding to the site of the fibrillar centre (Fig. 4). In a more frequent type of distribution, the silver grains were located between 'the site of

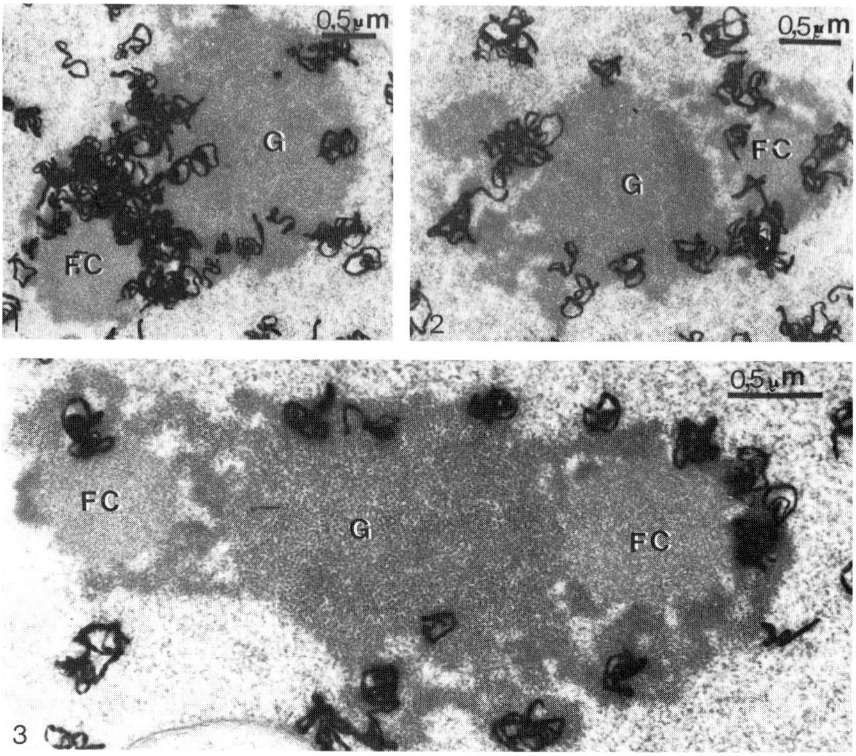

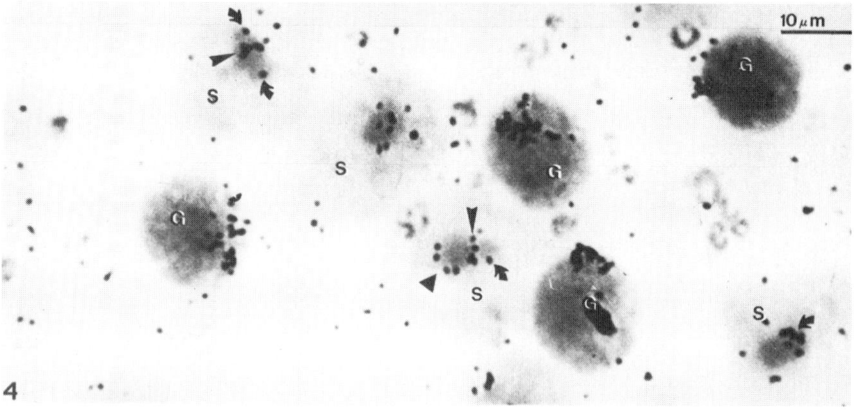

Figures 1-3. - Autoradiographs of human Sertoli cell nucleoli following ^{3}H-uridine incorporation. The labelling grains are located over the fibrillar strands surrounding the fibrillar centre(s) (Fig. 1, 2 and 3), between the fibrillar centre and the granular part(Fig. 1) and over the fibrillo-granular strands (Fig. 2 and 3). FC : fibrillar centre ; G : granular part. Figure 4. - In situ hybridization performed with 28S cDNA. S : Sertoli cell ; G : germinal cell. The ribosomal genes are located : 1. over and at the border of the spherule corresponding to the fibrillar centre (➔); 2. between this sperule and the main (granular) part (➤) ; 3. around the main part corresponding to the granular component (▶).

the fibrillar centre and that of the granular part, and also around the latter. Another type of distribution revealed a few grains at the periphery of the pink sperule corresponding to the fibrillar centre and a cluster of grains between this site and that of the granular part (Fig. 4).

The distribution of the label after ^{3}H-uridine incorporation corresponds to that described in other cell types (review in Stahl, 1982 and Goessens, 1984). However in our experiments, the real duration of ^{3}H-uridine incorporation by the Sertoli cells is not known with precision. It mainly depends on two factors : the rate of penetration of the labelled precursor and the position of the Sertoli cell in relation to the border of the incubated fragment. As it is generally accepted that the nascent pre-rRNA is not immediately detached from rDNA (review in Franke et al., 1979), a shift of the transcription product is unlikely. Accepting that the site of label corresponds to the site of rDNA transcription, there is a good correlation between the distribution of the label and the location of ribosomal genes revealed by in situ hybridization. If rDNA were strictly confined within the fibrillar centre and/or the surrounding layer of dense fibrils, only one or two clusters of grains should be seen following <u>in situ</u> hybridization, since the human Sertoli cell nucleolus contains one or two fibrillar centres. On the contrary, a dispersion of the grains is more often observed, which shows that the distribution of rDNA is not strictly superposed on fibrillar centres. The human Sertoli cell nucleolus is a new example of uncoiling of rDNA outside the NOR represented by the fibrillar centre.

AKNOWLEDGEMENTS

We gratefully acknowledge Dr J.L. Bergé-Lefranc for biochemical help. We thank C. Cataldo and M. Soler (CNRS) for the microphotographs and Mrs M. Gandriaut for the preparation of the manuscript.

This work was supported by CNRS (UA 1189) INSERM (Contrat n° 85 4 023), MRT (Contrat n° 83 C 0670) and Fondation pour la Recherche Médicale.

REFERENCES

Brahic M., Stroop W.G. and Baringer J.R., 1981. Theiler's virus persists in glial cell during demyelinating disease. Cell, 26, 123.

Devictor M., Hartung M. and Stahl A., 1984. Distribution of fibrillar centres and silver-stained components in the nucleolus of human Sertoli Cells. Biol. Cell, 50, 103-106.

Franke W.W., Scheer U., Spring H., Trendelenburg M.F. and Zentgraf H., 1979. Nucleolar chromatin. Definition and diversity of contents and arrangements of genes. In : "The cell nucleus" H. Busch ed., vol. 7, pp 49-95, Academic Press New York.

Gall J.G. and Pardue M.L., 1971. Nucleic acid hybridization in cytological preparations. Methods Enzymol., 21, 470-480.

Goessens G., 1984. Nucleolar structure. Int. Rev. Cytol., 87, 107-158.

Hernandez-Verdun D., Hubert J., Bourgeois C.A. and Bouteille M., 1980. Ultrastructural localization of Ag-NOR stained proteins in the nucleolus during the cell cycle and in other nucleolar structures. Chromosoma (Berl.), 79, 349-362.

Luciani J.M., Devictor-Vuillet M., Gagne R. and Stahl A., 1974. An air-drying method for first meiotic prophase preparations from mammalian ovaries. J. Reprod. Fertil., 36, 409-411.

Mirre C. and Knibiehler B., 1982. A re-evaluation of the relationships between the fibrillar centres and nucleolus-organizing regions in reticulated nucleoli : ultrastructural organization, number and distribution of the fibrillar centres in the nucleolus of the mouse Sertoli cell. J. Cell Sci., 55, 247-259.

Mirre C. and Stahl A., 1981. Ultrastructural organization, sites of transcription and distribution of fibrillar centres in the nucleolus of the mouse oocyte. J. Cell Sci., 48, 105, 126.

Scheer U. and Rose K.M., 1984. Localization of RNA polymerase I in interphase cells and mitotic chromosomes by light and electron microscopic immunocytochemistry. Proc. Natl. Acad. Sci. (USA), 81, 1431-1435.

Stahl A., 1982. The nucleolus and the nucleolar chromosomes. In : "The Nucleolus", ed. E.G. Jordan and C.A. Cullis, Cambridge University Press, pp. 1-24.

Stahl A., Mirre C., Hartung M. et Knibiehler B., 1980. Localisation, structure et activité des gènes ribosomiques dans le nucléole de l'ovocyte en prophase de méiose. Reprod. Nutr. et Dével., 20, 469-483.

Wachtler F., Ellinger A. and Schwarzacher H.G., 1980. Nucleolar changes in human phytohaemagglutinin-stimulated lymphocytes. Cell Tiss. Res., 213, 351-360.

Association of ribosomal genes in the human oocyte at meiotic prophase: cytogenetic consequences

E. Jullian, M. Hartung, A. M. Vagner-Capodano and A. Stahl

Laboratoire de Génétique, Faculté de Médecine, 27 boulevard Jean Moulin, 13385 Marseille Cédex 5, France

INTRODUCTION

The human acrocentric chromosomes, which contain a nucleolus organizer region (NOR) in their short arm, are often involved in numerical and structural anomalies. Trisomies of acrocentric chromosomes represent almost 40 % of all human lethal trisomies (Boué et al., 1975). In Down syndrome, the extra chromosome 21 originates in most cases as a result of nondisjunction during the first meiotic division of the oocyte (Mattei et al., 1979). The incidence of the Robertsonian translocations, which are the most frequent form of chromosome rearrangement in man, is about 1 ‰ in the general population. In 90 % of Robertsonian translocations, nonhomologous acrocentrics are involved, the most frequently observed being the t(13;14) and t(14;21). These facts have led Hartung et al. (1979), Mirre et al. (1980) and Stahl et al. (1983) to investigate whether a structural arrangement exists at prophase I of human meiosis that might favor the failure of disjunction of acrocentrics and the formation of Robertsonian translocations. Using NOR-silver techniques and electron microscopy they showed that NORs belonging to several chromosomes are associated at pachytene within the fibrillar centre of the nucleolus. Mirre et al. (1980) observed the penetration into the fibrillar centre of chromatin fibers emanating from 1 to 3 bivalents and postulated that the ribosomal genes from 4 to 12 chromatids were temporarily juxtaposed inside the same nucleolar component. However since morphological or staining techniques are not sufficiently specific, we have undertaken a study of the early stages of meiotic prophase using in situ hybridization for localizing the ribosomal genes in the human oocyte. The sites containing rDNA were compared to those revealed by the silver technique and the electron microscope.

MATERIAL AND METHODS

Ovarian material was recovered from ten fetuses, aged 16 to 24 weeks.

Electron microscopy.

The ovaries were cut into 1-mm^3 pieces and fixed with 3 % glutaraldehyde in 0.1 M phosphate buffer, pH 7.2/0.05 M sucrose at 4°C followed by 2 % osmium tetroxide in the same buffer. After dehydration in a graded series of acetone, the specimens were embedded in Epon. The sections showing silver interference color were stained with uranyl acetate and lead citrate.

Silver staining of the NOR.

Light microscope silver staining. After isolation and spreading using the technique of Luciani et al. (1974), the germ cells were stained according to the technique of Goodpasture and Bloom (1975).

Electron microscope silver staining. Silver impregnation was performed by the technique of Hernandez-Verdun et al. (1980).

In situ hybridization.

Preparation of the probe. A ^{3}H-DNA probe complementary to 28S rRNA was prepared according to the method proposed by Maniatis and al. (1982) using the Avian Myeloblastosis virus reverse transcriptase (Bethesda Research Laboratory).

The RNA was extracted from a culture of Hela cells according to Favaloro and al. (1980) modified by Maniatis and al. (1982). The poly $(A)^+$ RNA were separated by chromatography on an oligo (dT)-cellulose column. The 28S rRNA template was purified by a sucrose gradient (15-30 % W/V SW 65 Ti 3 hr at 48,000 rpm). ^{3}H dCTP (64 ci/mmol, Amersham) was used as labelled nucleotide.

The cDNA probe was purified on a Sephadex G-50 column and then ethanol precipitated. The specific radioactivity was 8. 10^7 dpm/µg.

Hybridization. Hybridization was performed according to the general technique discribed by Gall and Pardue (1971).

The slides were pretreated by RNAse (100 μg/ml in 2XSSC) at 37°C at least 1 hr then washed in 2XSSC, dehydrated in graded ethanol and air dried.

Chromosomal DNA was denatured in 70 % formamide-2XSSC for 2 min at 70°C. Then the slides were quickly cooled in 70 % ethanol-at -20°C, then dehydrated and air dried.

The hybridization mixture contained 50 % desionised formamide V/V, 2XSSC, 10 % dextran sulphate and 2 μg/ml of sheared, denatured salmon sperm DNA.

10 μl of the hybridization mixture with 2ng of the probe, were used per slide covered with an acid treated, siliconized coverslip. The coverslips were sealed with rubber cement. The slides were incubated overnight in a moist chamber at 37°C.

The removal of non specific bound probe was performed first in 2XSSC-50 % formamide at 39OC and then several times in 2XSSC at 39OC.

The slides were dehydrated in ethanol and air dried.

The slides were dipped at 45°C in Kodak NTB2 emulsion diluted 1:1 with distilled water. After at least 2 hr drying, slides were put in light tight plastic boxes with silicagel and exposed 7-10 days at 4°C.

The slides were developped in Kodak D-19 and stained with Giemsa.

RESULTS AND DISCUSSION

The number and the aspect of nucleoli and NORs showed variations from leptotene to pachytene. At leptotene, the nucleus contained 3-6 different-sized nucleoli, silver-stained NORs were localised in the peripheral region of the nucleoli. The average number of Ag-positive structures was seven. At pachytene, only 1-3 nucleoli were present. Often, each nucleolus was connected with several acrocentric bivalents. The nucleolus was divided into two zones. One of them was silver-positive and corresponded to the NOR. The average number of Ag-positive structures decreased to 2.5 per nucleus. The short arm of the acrocentric bivalents was connected with the silver-stained area of the nucleolus. Often, two acrocentric bivalents were associated with the Ag-positive area (fig. 1).

The sites of rDNA revealed by *in situ* hybridization corresponded to those shown by the silver technique : they displayed the same location, number and variation. At leptotene, the autoradiographs showed 1-3 grain clusters at the border of the nucleoli. This location was similar to that of silver-stained NORs. At zygotene with typical bouquet polarisation, well separated clusters of grains were localised in the area where the telomeres of the bivalents converged. At pachytene, in the main nucleolus gathering of all labelling grains generally occurred in an area where the short arm of several acrocentric bivalents converged (fig. 2 and 5). In the electron microscope, the nucleolus showed a segregation of its components. The fibrillar centre was separated from the fibrillar and granular components. The layer of dense fibrils that usually surrounds the fibrillar centre and corresponds to the zone of rDNA transcription was absent. The short arm of the acrocentric bivalents was associated with the fibrillar centre (fig. 3). Chromatin fibres emanating from the secondary constriction region penetrated into the fibrillar centre. The number of acrocentric bivalents connected with the fibrillar centre varied from one to three.

At early diplotene, the chromosomes forming the bivalents were separated along their length except at chiasmata. The short arm of acrocentric bivalents displayed two separated filaments that were tied to the nucleolar surface. The nucleus contained two to four nucleoli. They had either the same size, either different sizes and in that case we could distinguish one large-sized main nucleolus and one to three small secondary nucleoli.

In situ hybridization autoradiographs showed 5-7 clusters of silver grains, whatever the number of nucleoli. They were consistently located at the border of the nucleoli (fig. 4 and 5). The small secondary nucleoli displayed only one grain cluster. The main nucleoli showed 3 to 5 clusters. In well spread nuclei, the clusters were located at the end of a chromosome applied to the nucleolar surface (fig. 5, arrow).

In the electron microscope, the nucleoli did not display any more the spontaneous segregation that characterized the pachytene

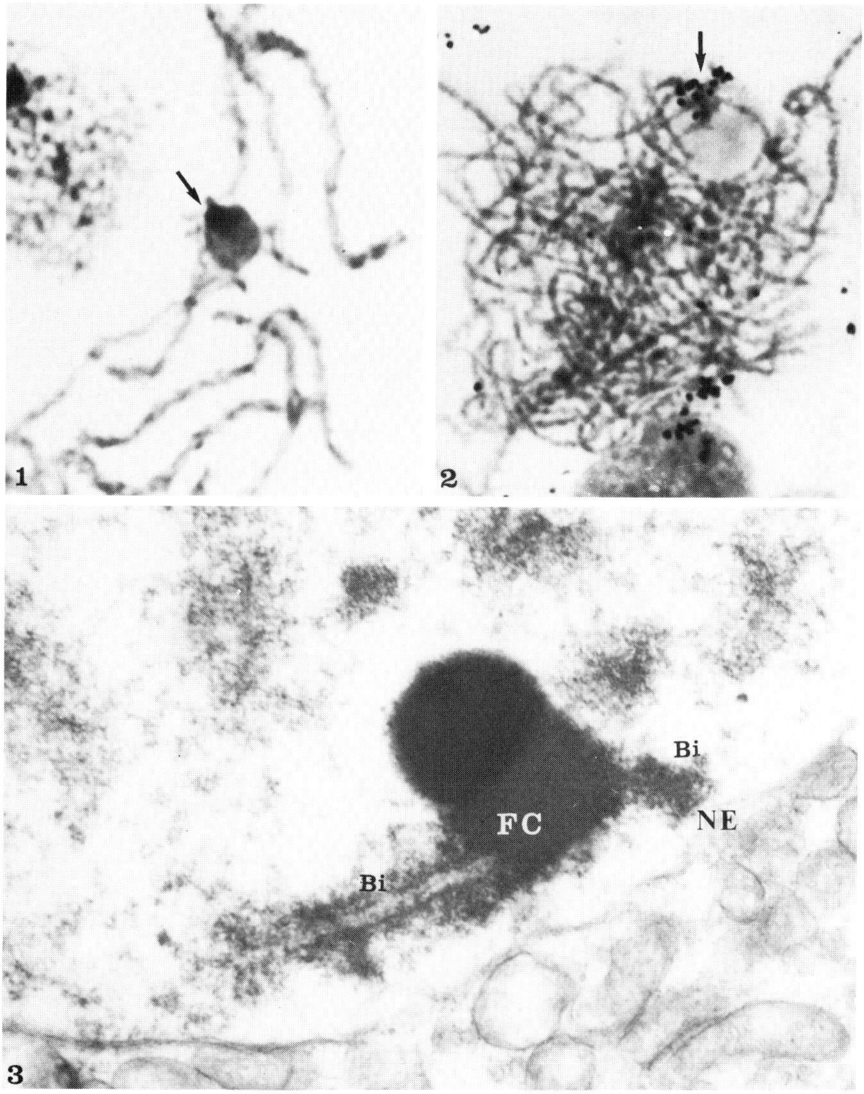

Fig. 1-3. Pachytene stage oocytes from a 18 week-old fetus.

Fig. 1. Detail of a pachytene nucleus stained by the NOR-silver technique. Two acrocentric bivalents are associated with the Ag-positive area of the nucleolus (arrow) x 4,000.
Fig. 2. Autoradiograph after *in situ* hybridization with 28S (^{3}H) cDNA. The grain cluster (arrow) is located in the nucleolar area corresponding to the silver-stained NOR in fig. 1. x 3,200.
Fig. 3. The fibrillar centre (FC) of the nucleolus is connected with two bivalents (Bi), one of them showing the synaptonemal complex. NE : nuclear envelope. x 12,000.

stage. It had a reticulated (nucleolonemal) structure. Fibrillar centres were localised at the nucleolar periphery. They were partially surrounded by a layer of dense fibrils (fig. 6). The silver technique performed at the ultrastructural level showed that the fibrillar centres and the surrounding dense fibrils were silver-positive (fig. 7).

Comparison of the ultrastructural nucleolar aspects with the results of silver staining and with the labelling sites of *in situ* hybridization suggests that fibrillar centres can be identified with nucleolus organizers. The data which favor that homology are discussed by Schwarzacher and Wachtler, Goessens, Thiry and Lepoint, Jordan, Scheer and Raska in this volume. When comparing the results of *in situ* hybridization to those of morphological techniques, two situations must be distinguished : 1. If the fibrillar centre is surrounded by a layer of dense fibrils, the clustered labelling indicates that rDNA is located either inside the fibrillar centre, or within the shell of dense fibrils, or in both components. In active nucleoli, the two components are usually joined forming a functional unit (Stahl et al., 1977). 2. Transcription is temporarily put off in the human oocyte at pachytene (Hartung and Stahl, 1978). At this stage the fibrillar centre is not surrounded by dense fibrils (Mirre and Stahl, 1980 and this paper, fig. 3). Thus, at pachytene the labelling sites of *in situ* hybridization correspond only to the fibrillar centre, as already described in the human spermatocyte at pachytene (Arroua et al., 1982).

Our observations demonstrate that the ribosomal genes, which form well individualized NORs at leptotene, gather and associate at pachytene. They confirm and extend the previous descriptions of Hartung et al. (1979) and of Mirre et al. (1980), which were based on morphological criteria.

Fig. 4. Early diplotene from a 20 week-old fetus. *In situ* hybridization with 28S (^{3}H)cDNA. The main nucleolus (large arrow) and the secondary nucleolus (small arrow) show respectively five and one labelled sites in their peripheral region. x 2,800.

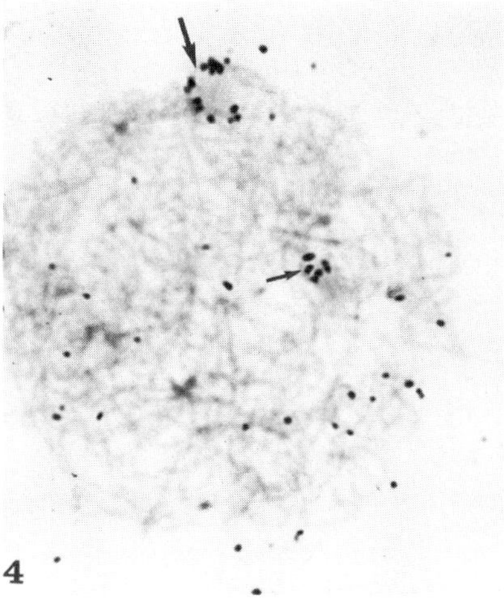

4

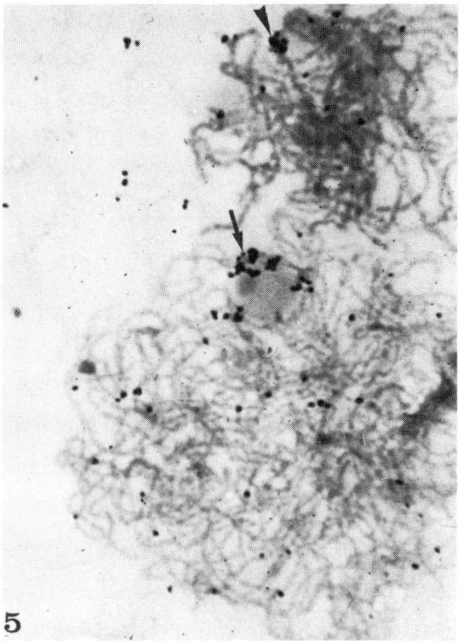

5

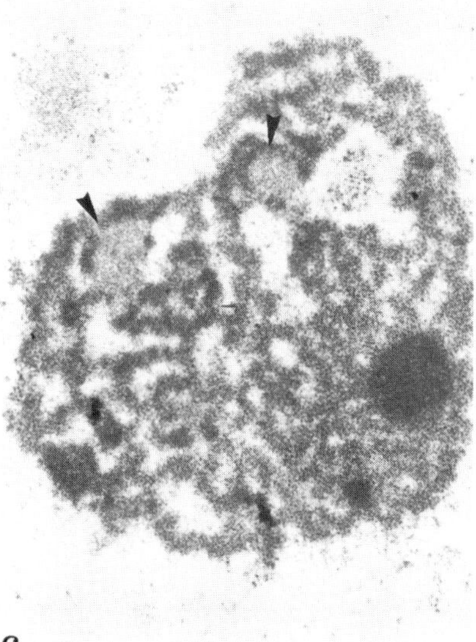

6

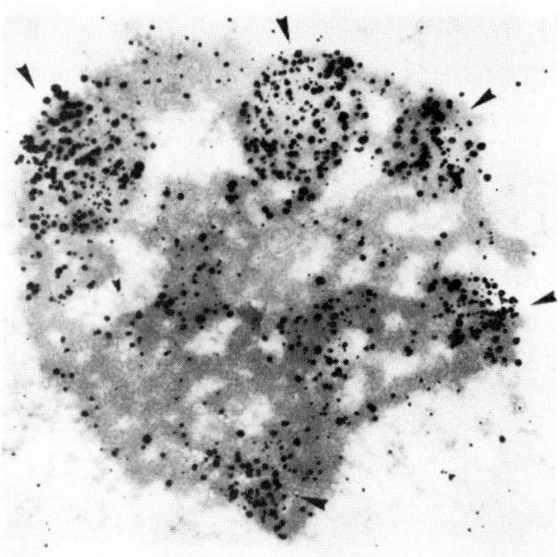

7

Fig. 5. Early diplotene and pachytene nuclei from a 20 week-old fetus. *In situ* hybridization with 28S (^{3}H)cDNA. The diplotene nucleolus (arrow) shows 5 grain clusters in its peripheral region. A clustered labelling (arrowhead) is located at the border of the pachytene nucleolus. x 2,800.

Fig. 6. Electron micrograph of an oocyte nucleolus at diplotene from a 24 week-old fetus, showing peripheral fibrillar centres partially surrounded by dense fibrils (arrowheads). x 15,000.

Fig. 7. Electron micrograph of an oocyte nucleolus at diplotene from a 24 week-old fetus, stained with silver. The peripherally located fibrillar centres and the surrounding dense fibrillar layer are Ag-positive. x 15, 000.

The average number of NORs identified by silver staining or *in situ* hybridization was 7 at leptotene, 2.5 at pachytene and 6 at early diplotene. The number found at early diplotene was lower than the one observed at more advanced diplotene by Hartung et al. (1979). However, these numbers illustrate the fact that the NORs associated at pachytene begin to separate as soon as the oocyte enters the diplotene stage. At early diplotene, the two short arms of acrocentric bivalents are still in close proximity (Vagner-Capodano et al., 1987) so that two homologous NORs can be contained in a single silver-positive area or can give a single "hot-spot" labelling with *in situ* hybridization. The association of ribosomal genes within the nucleolar fibrillar centre at pachytene results in bringing together chromatids in the short arm region of acrocentric bivalents. This observation is in agreement with tridimensional reconstructions of pachytene nuclei made by Bojko (1983). Electron microscope studies show that tight contact of chromatids is formed by overlapped or intermingled lateral loops. A breakage-reunion event between nonhomologous chromatids may result in a Robertsonian translocation (Stahl et al., 1983). The accidental recombination between nonhomologous requires formation of an inverted loop on one chromatid (Guichaoua et al., 1986).

It has been suggested that the presence of ribosomal genes from two paired chromosomes inside the same fibrillar centre may also be a causal factor inducing nondisjunction (Mirre et al.,1980). Whether the argyrophilic protein of the fibrillar centre in which the ribosomal genes are embedded may hinder the separation at metaphase I of the two paired chromosomes is still uncertain. However, our findings support the suggestion of Ferguson-Smith and Handmaker (1961, 1964) and of Evans (1967) that the acrocentric chromosomes are more prone to nondisjunction because they are often associated with a common nucleolus.

AKNOWLEDGEMENTS

We thank C. Cataldo (CNRS), A. de Lanversin and A. Calisti for technical assistance, M. Soler (CNRS) for the microphotographs and Mrs M. Gandriaut for the preparation of the manuscript. we also thank Prof. R.D. Schmickel (Human Genetics Dpt-University of Pennsylvania) for his kind invitation in his Laboratory.

This work was supported by CNRS (UA 1189) INSERM (contrat n° 85 4023) and Fondation pour la Recherche Médicale.

REFERENCES

Arroua, M.L., Hartung, M., Devictor, M., Bergé-Lefranc, J.L., Stahl, A. 1982. Localisation of ribosomal genes by *in situ* hybridization in the fibrillar centre of the nucleolus in the human spermatocyte. Biol. Cell 44, 337-340.

Bojko, M. 1983. Human meiosis VIII. Chromosome pairing and formation of the synaptonemal complex in oocytes. Carlsberg Res. Commun. 48, 457-483.

Boué, J., Boué, A., Lazar, P. 1975. Retrospective and propective epidemiological studies of 1,500 karyotypes of spontaneous human abortions. Teratology 12, 11-26.

Evans, H.J. 1967. The nucleolus, virus infection and trisomy in man. Nature (London) 214, 361-363.

Hernandez-Verdun, D., Hubert, J., Bourgeois, C.A., Bouteille, M. 1980. Ultrastructural localization of Ag-NOR stained proteins in the nucleolus during the cell cycle and in other nucleolar structures. Chromosoma 79, 349-362

Favaloro, J., Treisman, R., Kamen, R. 1980. Transcription maps of polyoma virus-specific RNA : Analysis by two-dimensional nuclease S1 gel mapping. Methods Enzymol. 65, 718.

Ferguson-Smith, M.A. 1964. The sites of nucleolus formation in human pachytene chromosomes. Cytogenetics 3, 124-134.

Ferguson-Smith, M.A., Handmaker, S.D. 1961. Observations on the satellited human chromosomes. Lancet i, 638.

Gall, J.G., Pardue, M.L. 1971. Nucleic acid hybridization in cytological preparations. Methods Enzymol. 21, 470-480.

Goodpasture, C., Bloom, S.E. 1975. Vizualization of nucleolar organizer regions in mammalian chromosomes using silver staining. Chromosoma 53, 37-50.

Guichaoua, M.R., Devictor, M., Hartung, M., Luciani, J.M., Stahl, A. 1986. Random acrocentric bivalent associations in human pachytene spermatocytes. Molecular implications in the occurrence of Robertsonian translocations. Cytogenet. Cell Genet. 42, 191-197.

Hartung, M., Stahl, A. 1978. Autoradiographic study of RNA synthesis during meiotic prophase in the human oocyte. Cytogenet. Cell Genet. 20, 51-58.

Hartung, M., Mirre, C., Stahl, A. 1979. Nucleolar organizers in human oocytes at meiotic prophase I, studied by the silver-NOR method and electron microscopy. Hum. Genet. 52, 295-308.

Luciani, J.M., Devictor-Vuillet, M., Gagné, R., Stahl, A. 1974. An air drying method for first meiotic prophase preparations from mammalian ovaries. J. Reprod. Fert. 36, 409-411.

Maniatis, T., Fritsch, E.F., Sambrook, J. 1982. Molecular cloning. A laboratory manual. Cold Spring Harbor Laboratory Press New York.

Mattei, M.G., Mattei, J.F., Ayme, S., Giraud, F. 1979. Origin of the extra chromosome in trisomy 21 children. Human Genetics 46, 107-110.

Mirre, C., Hartung, M., Stahl, A. 1980. Association of ribosomal genes in the fibrillar center of the nucleolus : a factor influencing translocation and non-disjunction in the human meiotic oocyte. Proc. Natl. Acad. Sci. USA 77, 10, 6017-6021.

Stahl, A., Mirre, C., Hartung, M., Knibiehler, B., Navarro, A. 1977. Localization, structure and activity of ribosomal cistrons in the mouse oocyte during meiotic prophase I. Chromosomes Today Vol. 6, A. de la Chapelle and M. Sorsa eds. 255-264.

Stahl, A., Luciani, J.M., Hartung, M., Devictor, M., Bergé-Lefranc J.L., Guichaoua, M.R. 1983. Structural basis for Robertsonian translocations in man : Association of ribosomal genes in the nucleolar fibrillar center in meiotic spermatocytes and oocytes. Proc. Nat. Acad. Sci. (USA) 80, 5946-5950.

Vagner-Capodano, A.M., Hartung, M., Stahl, A. 1987. Nucleolus, nucleolar chromosomes and nucleolus-associated chromatin from early diplotene to dictyotene in human oocyte. Hum. Genet. (in press).

Author index

Subject index